Advances in Applied Mechanics

Volume 12

Editorial Board

Contributors to Volume 12

P. Germain

J. F. Harper

V. C. Liu

Theodore H. H. Pian

Pin Tong

ADVANCES IN
APPLIED MECHANICS

Edited by Chia-Shun Yih

COLLEGE OF ENGINEERING
THE UNIVERSITY OF MICHIGAN
ANN ARBOR, MICHIGAN

VOLUME 12

1972

ACADEMIC PRESS New York and London

ACADEMIC PRESS, INC.
111 Fifth Avenue, New York, New York 10003

United Kingdom Edition published by
ACADEMIC PRESS, INC. (LONDON) LTD.
24/28 Oval Road, London NW1

LIBRARY OF CONGRESS CATALOG CARD NUMBER: 48-8503

PRINTED IN THE UNITED STATES OF AMERICA

Contents

List of Contributors

Numbers in parentheses indicate the pages on which the authors' contributions begin.

P. GERMAIN, University of Paris (VI), Paris, France (131)

J. F. HARPER, Department of Mathematics, Victoria University of Wellington, Wellington, New Zealand (59)

V. C. LIU, The University of Michigan, Ann Arbor, Michigan (195)

THEODORE H. H. PIAN, Department of Aeronautics and Astronautics, Massachusetts Institute of Technology, Cambridge, Massachusetts (1)

PIN TONG, Department of Aeronautics and Astronautics, Massachusetts Institute of Technology, Cambridge, Massachusetts (1)

Preface

In mechanics, as in any other branch of physical science, the initial advances must of necessity be made in the discovery of physical laws, often through observation and experimentation, and in their mathematical formulation. After this is done, further efforts can be made in many directions.

Theories based on these laws may be developed that continually enrich the field governed by these laws. Professor Germain's article in this volume represents an excellent effort in this direction. Or methods of calculation may be devised to apply the physical laws discovered to special problems. The authoritative article by Professors Pian and Tong falls in this category. Incidentally, the latter article also illustrates how the manner and direction of research in a field can be profoundly influenced by the availability of a new tool—in this case the modern computer.

Along with these efforts, human desire to understand nature is of course the motivation for many theoretical, experimental, or observational studies in fields which are either of practical importance or of great intellectual interest. Much is to be done in these fields even though the basic equations governing the phenomena are known. One such field is that of two-phase flow, one aspect of which is treated in Dr. Harper's informative and interesting article. Another is that of interplanetary gas dynamics, treated comprehensively by Professor Liu. The phenomena in the two fields differ tremendously in scale, and their concatenation evokes in fluid dynamicists a sense of wonder of the scope of their domain of cultivation.

Thus the four articles in this volume show that in the various aspects of mechanics research, advances are not always made in sequence or according to plans, but are often made simultaneously and independently, complementing and reinforcing one another in the progress of their common front.

CHIA-SHUN YIH

Finite Element Methods in Continuum Mechanics

THEODORE H. H. PIAN AND PIN TONG

Department of Aeronautics and Astronautics
Massachusetts Institute of Technology, Cambridge, Massachusetts

I. Introduction

The finite element method is an approximate method for solving a field problem. In this method a domain V is decomposed into a finite number of nonoverlapping subdomains V_n which are called the elements. The procedure is first to seek an approximate solution within each element and to characterize the behavior of the element by a finite number of unknown parameters. A suitable procedure is then employed to combine the relations for the individual elements into a system of equations to be used to solve these unknown parameters. When the element size becomes smaller and smaller the discretization errors of the field variables vanish and the exact solution can be obtained. The unknown parameters are usually the value of field variables at a finite number of points which are called nodes, and which may be on the boundary of or within the elements. The approximate solution can simply be obtained by interpolating these field variables within each element. The characterization of the element and the establishing of the system of equations for the unknown parameters are usually based on the stationary condition of a functional based on a certain variational principle or a variational statement. The variational formulation is equivalent to the original field equations. When the variational formulation is based on the minimum principle, the unknown parameters may also be determined by a systematic mathematical searching technique instead of the solution of a system of equations.

The finite element method was first introduced in solid mechanics applications in the mid 1950s as an extension of the so-called matrix methods in structural analysis (Argyris, 1958, 1960, Livesley, 1964). During the decade 1945–1955 intensive advances were made in systematic methods for analyzing complex structures which may contain large number of components. Some typical structural components, for example, are simple bars in tension or compression and beam elements under bending and twisting. They are connected at a finite number of nodal points. The fundamental idea of the matrix methods in structural analysis is to utilize the relations between the displacements and internal forces at the nodal points of the individual structural components to form a system of algebraic equations with the nodal displacements or the nodal internal forces, or sometimes both nodal displacements and forces as unknowns. The corresponding methods are generally classified as the displacement method, the force method, or the mixed method. The development of the equations for discrete structural systems is in many aspects similar to the setup of hydraulic and electrical networks: hence is quite familiar to the engineers. The derivation of the equations can be carried out most conveniently by using the matrix notation. The solution of the equations and, in some instances, many operations in the derivation of such equations are carried out by high speed digital computers.

The term "finite element method" was first introduced by Clough (1960). A few years earlier, Turner *et al.* (1956), in applications to aircraft structural analysis, extended the matrix displacement method to plane stress problems by using triangular and rectangular elements. In this formulation the behavior of each element is represented by an element stiffness matrix which relates the forces at the finite number of nodal points of the element to the nodal displacements. In contrast to the conventional matrix structural analysis for which the relation between forces and displacements for each structural component is derived exactly, the solution of the plane stress problems is based only on approximate displacement functions within each element. The comprehensive treatise of the energy theorems and matrix methods by Argyris (1960), in fact, already provided a pioneering effort in deriving the element stiffness matrix of a plane stress rectangular panel.

It was recognized later that such a finite element method is a generalized Ritz method based on the principle of virtual work or the principal of minimum potential energy. Historically speaking, Courant (1943) already presented an approximate solution of the St. Venant torsion problem by assuming a linear distribution of the stress function in each of the assemblage of triangular elements. The finite element method, however, is a much more versatile method in comparison with the conventional Ritz method. In the Ritz method the assumed displacement modes extended to the entire domain; thus, for a solid continuum of irregular boundary geometry the choice of the admissible displacement modes would be very difficult. Also, in that case for an accurate solution a very large number of assumed modes must be used and considerable algebraic manipulation must be performed. Furthermore, even for the same type of solid continuum in general, if the boundary conditions are different a new set of assumed displacement modes must be used. For the finite element method, however, the same assumed displacement functions which are usually very simple can be used for the individual element, since the accuracy of the solution now depends on the number of elements employed. In solving problems of the same type of solid continuum but of different configurations by the finite element method it is sometimes only necessary to alter a few input cards in the digital computation. For example, a problem which involves multiple-connected domain will not be more difficult than simply-connected ones, and with only a slight modification a computer program which has been used for isotropic materials can be used for general anisotropic materials.

For certain boundary value problems which involve regular shaped boundaries, and hence regular mesh arrangements can be used in the finite element formulation, the resulting equations are sometimes identical to that of the conventional finite difference formulation (Pian, 1971b). The finite element method, however, always has its advantage in the application to irregular domains and to nonhomogeneous mediums.

During the last decade the formulations of finite element methods by different types of variational principles in elasticity were discussed by Besseling (1963), Melosh (1963), Jones (1964), Gallagher (1964), Pian (1964), Fraeijs de Veubeke (1964), Herrmann (1965), Prager (1967, 1968), Tong and Pian (1969), and Tong (1970). Indeed, because the assumed functions for the field variables are only piecewise continuous, new and modified variational principles have been derived to allow for the discontinuity of the field variables at the interelement boundaries. One of the main objectives of the present article is to introduce the various finite element formulations for linear elastic solids based on the variational approaches.

The formulations of the finite element methods are, of course, not restricted to the variational approach. Oden (1969) formulated the equations for the finite element analysis of the thermoelasticity problems using the so-called energy balance method. Szabo and Lee (1969) formulated the finite element solution of plane elasticity using the Galerkin method. Indeed, the interpolation functions have been used by mathematicians for solving partial differential equations of the boundary value problems and the convergence characteristics have been carefully examined (Babuska, 1971; Birkhoff *et al.*, 1968; Fix and Strang, 1969; Zlamal, 1968).

The finite element methods have also been extended to other field problems in continuum mechanics. For example, the twisting of prismatic bars, the steady state heat conduction, the potential flow of ideal fluids, and a number of other continuum mechanics problems are governed by similar elliptic equations, and hence can all be solved by the same finite element scheme. It turns out that most of the existing finite element solutions for fluid mechanics and diffusion problems are either based on certain variational principles or can be expressed in the form of variational statements. Thus, the experiences that have been obtained in the finite element solution of solid mechanics problems can be applied to some of the other continuum mechanics problems. For a certain boundary value problem in continuum mechanics there also exists a way to convert the solutions into integral equations with variables defined over the surface of the domain. A finite element method can be formulated for this derived problem by subdividing the surface into discrete elements.

In the short duration of a decade and a half since the concept of the finite element method was introduced to applied mechanicians, the research and development works in the field have, indeed, mushroomed. Even the various recently published review acticles (Felippa and Clough, 1970; Zienkiewicz, 1970; Zudans, 1969) and the revised text by Zienkiewicz (1971) do not have complete coverage of this subject. Supplementary materials must be obtained by going over the proceedings of several national and international conferences on matrix and finite element methods (Przemieniecki *et al.*, 1966; Berke *et al.*, 1969; Holand and Bell, 1969; Rowan and Hackett, 1969; Sörensen, 1969; Gallagher *et al.*, 1971; Fraeijs de Veubeke, 1971), and by searching through the latest professional journals in solid, structural, and fluid mechanics, in heat transfer, and in numerical methods. The developments in finite element methods have, indeed, extended from the original aeronautical structural applications to the field of civil, mechanical, naval architectural, and nuclear engineering.

Obviously it is impossible for the authors to present a complete survey of the subject of the finite element method in a relatively brief contribution. The present article is thus restricted to a brief introduction of the various finite element formulations for linear elastic solids and to discuss similar formulations for several other field problems. These are given respectively in Sections II and III. In these sections detailed illustrations will be presented for several typical finite element formulations. In the Appendix a brief account is given of the interpolation functions which are needed in all finite element formulations. Material has been drawn from several papers by the authors (Pian and Tong, 1969b; Pian, 1970, 1971a; Tong, 1971) in the preparation of the present article.

II. Finite Element Formulation of Linear Elasticity Problems

A. General Discussion

The linear elasticity problems are governed by three sets of equations in terms of rectangular Cartesian coordinates,

1. Stress equilibrium equations

$$\sigma_{ij,j} + \bar{F}_i = 0, \qquad i = 1, 2, 3. \tag{2.1}$$

2. Stress–strain relations

$$\sigma_{ij} = C_{ijkl}\,\varepsilon_{kl} \tag{2.2}$$

or

$$\varepsilon_{ij} = S_{ijkl}\,\sigma_{kl}. \tag{2.3}$$

3. Strain–displacement relation

$$\varepsilon_{ij} = \tfrac{1}{2}(u_{i,j} + u_{j,i}), \tag{2.4}$$

where

σ_{ij} = stress tensor component,
ε_{ij} = strain tensor component,
$\bar{F}_i$ = prescribed body force component,
C_{ijkl} = elastic stiffness coefficient,
S_{ijkl} = elastic compliance coefficient.

An elasticity problem is to solve (2.1), (2.2), and (2.4) for the domain V when along the boundary S_σ the surface tractions T_i are prescribed and along the remaining boundary S_u the displacements u_i are prescribed. The surface tractions T_i are related to the stresses by

$$\sigma_{ij}\nu_j = T_i, \tag{2.5}$$

where ν_j is the direction cosine of the surface normal. This equation may be interpreted as the equilibrium condition between the stresses and the surface tractions.

Alternate ways of expressing the elasticity equations are the variational formulations. The three commonly used variational principles for small displacement theory of elasticity are

(a) Principle of minimum potential energy, which can be derived directly from the principle of virtual work and for which the only field variables, the displacements, must be continuous within the domain.

(b) Principle of minimum complementary energy for which the only field variables, the stresses, must be in equilibrium.

(c) Reissner's Variational Principle which has both displacements and stresses as the field variables.

The derivation of one of these principles from another can be accomplished through the introduction of appropriate conditions of constraint and the corresponding Lagrangian multipliers. One of the important applications of the variational principles is the Ritz method (or sometimes referred to as the Rayleigh-Ritz method) by means of which approximate functions are assumed for the field variables. Detailed discussions of variational methods in elasticity can be found from the text by Washizu (1968).

In the ordinary formulation of the Ritz method it is usually required that the assumed functions for the field variables should be continuous over the entire domain and should possess derivatives which are continuous up to the highest order occurring in the corresponding Euler differential equation of the variational problem. In the finite element formulation, displacement and/or stress fields are assumed to be continuous within each discrete element, but the continuity or equilibrium conditions along the interelement boundaries are to be relaxed to the extent that they are satisfied in an integral sense, and hence will be completely satisfied when the element size becomes infinitesimally small. This formulation thus calls for modified variational principles for which the continuity or equilibrium conditions along the interelement boundaries are introduced as conditions of constraint and appropriate boundary variables are used as the corresponding Lagrangian multipliers. Another flexibility in the finite element formulation in comparison to the conventional Ritz method is the broadening of the admissibility on the interelement boundary conditions to the degree that the functions shall possess continuous derivatives in such a manner that in addition to the fulfillment of the appropriate subsidiary conditions the functional of the variational problem is defined.

This section presents the various variational principles and the corresponding models used in the finite element formulation.

B. Principle of Minimum Potential Energy and the Compatible Model

The principle of minimum potential energy may be stated as the vanishing of the variation of the total potential energy functional π_p, which, for a solid being subdivided into a finite number discrete elements, may be written as

$$\pi_p = \sum_n \left[\int_{V_n} (\tfrac{1}{2} C_{ijkl} \varepsilon_{ij} \varepsilon_{kl} - \bar{F}_i u_i) \, dV - \int_{S_{\sigma n}} \bar{T}_i u_i \, dS \right] \tag{2.6}$$

where V_n is the volume of the nth element and $S_{\sigma n}$ is the portion of the boundary of the nth element over which the surface traction $\bar{T}_i$ is prescribed. In applying this principle, ε_{ij} is written in terms of the derivatives of the displacements u_i by (2.4) and the displacement functions should satisfy the continuity conditions. The differential equations and boundary conditions derived from this variational principle are obviously the equilibrium conditions given by (2.1) and (2.5) when the stresses are expressed in terms of displacements u_i.

In the finite element formulation the functions u_i are represented approximately over each element, say the nth element, by interpolation functions (see Appendix) and undetermined parameters q_n which are the values of the displacements and sometimes also derivatives of the displacements at a finite number of nodal points of the elements, Thus, q_n are termed the generalized nodal displacements. The interpolation functions are such that when the displacements at the nodes along the interelement boundary of two neighboring elements are compatible, the displacements along the corresponding interelement boundaries are also compatible. In matrix form the assumed displacements for the nth element may be expressed as

$$\mathbf{u} = \mathbf{f}\mathbf{q}_n, \tag{2.7}$$

where $\mathbf{q}_n$ is a column matrix of the element generalized displacements and $\mathbf{f}$ is a matrix of interpolation functions.

The stress–strain relation (2.1) may be written in matrix form,

$$\boldsymbol{\sigma} = \mathbf{C}\boldsymbol{\varepsilon}, \tag{2.8}$$

where

$$\boldsymbol{\sigma} = \{\sigma_{11} \quad \sigma_{22} \quad \sigma_{33} \quad \sigma_{12} \quad \sigma_{23} \quad \sigma_{31}\}, \tag{2.9}$$

$$\boldsymbol{\varepsilon} = \{\varepsilon_{11} \quad \varepsilon_{22} \quad \varepsilon_{33} \quad 2\varepsilon_{12} \quad 2\varepsilon_{23} \quad 2\varepsilon_{31}\}, \tag{2.10}$$

and $\mathbf{C}$ is the elastic constant matrix which is symmetric and is relating the stresses and the engineering strains. The strain displacement relation (2.8) may be written as

$$\boldsymbol{\varepsilon} = \mathbf{D}\mathbf{u}, \tag{2.11}$$

where $\mathbf{D}$ is defined as

$$\mathbf{D} = \begin{bmatrix} \partial/\partial x_1 & 0 & 0 \\ 0 & \partial/\partial x_2 & 0 \\ 0 & 0 & \partial/\partial x_3 \\ \partial/\partial x_2 & \partial/\partial x_1 & 0 \\ 0 & \partial/\partial x_3 & \partial/\partial x_2 \\ \partial/\partial x_3 & 0 & \partial/\partial x_1 \end{bmatrix}. \tag{2.12}$$

Substituting (2.7) and (2.11) into (2.6) we obtain

$$\pi_{\mathrm{p}} = \sum_{n=1}^{N} (\tfrac{1}{2}\mathbf{q}_n{}^T\mathbf{k}_n\mathbf{q}_n - \mathbf{q}_n{}^T\bar{\mathbf{Q}}_n), \tag{2.13}$$

where

$$\mathbf{k}_n = \int_{V_n} (\mathbf{Df})^T\mathbf{C}(\mathbf{Df})\,dV, \tag{2.14}$$

$$\bar{Q}_n = \int_{V_n} \mathbf{f}^T\bar{\mathbf{F}}\,dV + \int_{S_{\sigma n}} \mathbf{f}^T\bar{\mathbf{T}}\,dS, \tag{2.15}$$

and N is the total number of elements. In the above expression, $\bar{\mathbf{F}}$ and $\bar{\mathbf{T}}$ are respectively the prescribed body and surface forces. Since for each element $\frac{1}{2}\mathbf{q}_n{}^T\mathbf{k}_n\mathbf{q}_n$ is the strain energy in terms of the generalized nodal displacements, $\mathbf{k}_n$ is by definition the element stiffness matrix. The matrix $\bar{\mathbf{Q}}_n$ is a vector of the generalized forces due to the prescribed loads. Such generalized forces are equivalent nodal forces defined from the potential energy and are consistent with the assumed displacements.

The individual element stiffness matrix is derived directly by an approximate solution within each element based on the interpolation of the displacement functions in terms of the generalized nodal displacements. The element stiffness matrix characterizes the behavior of the element because, when there is only one isolated element, the equation

$$\mathbf{k}_n\mathbf{q}_n = \bar{\mathbf{Q}}_n \tag{2.16}$$

is an approximate relation between the generalized nodal displacements, $\mathbf{q}_n$, and the generalized nodal forces, $\mathbf{Q}_n$.

To fulfill the compatibility of the interelement nodal displacements a transformation matrix $\mathbf{J}$ can be introduced to relate the element nodal displacements $\mathbf{q}_n$ of the individual elements to a column of global displacements $\mathbf{q}$, the components of which being the generalized displacements of all the nodes in the entire domain,

$$\{\mathbf{q}_1\ \mathbf{q}_2 \cdots \mathbf{q}_N\} = \mathbf{J}\mathbf{q}. \tag{2.17}$$

The expression for π_p can thus be written as

$$\pi_\mathrm{p} = \tfrac{1}{2}\mathbf{q}^T\mathbf{K}\mathbf{q} - \mathbf{q}^T\bar{\mathbf{Q}}, \tag{2.18}$$

where*

$$\begin{aligned}\mathbf{K} &= \mathbf{J}^T[\mathbf{k}_1\ \mathbf{k}_2 \cdots \mathbf{k}_N]\mathbf{J},\\ \bar{\mathbf{Q}} &= \mathbf{J}^T\{\bar{\mathbf{Q}}_1\ \bar{\mathbf{Q}}_2 \cdots \bar{\mathbf{Q}}_N\}\mathbf{J},\end{aligned} \tag{2.19}$$

are respectively the stiffness matrix of the assembled structure and the column matrix of the applied generalized nodal forces.

Both the element stiffness matrix $\mathbf{k}_n$ and the assembled stiffness matrix $\mathbf{K}$ are positive semidefinite. However, if some of the generalized displacements are prescribed such that the remaining part of the partitioned matrix is positive-definite, the expression for π_p is

$$\pi_\mathrm{p} = \tfrac{1}{2}[\mathbf{q}_\alpha{}^T\ \bar{\mathbf{q}}_\beta{}^T]\begin{bmatrix}\mathbf{K}_{\alpha\alpha} & \mathbf{K}_{\alpha\beta}\\ \mathbf{K}_{\alpha\beta} & \mathbf{K}_{\beta\beta}\end{bmatrix}\begin{Bmatrix}\mathbf{q}_\alpha\\ \bar{\mathbf{q}}_\beta\end{Bmatrix} - [\mathbf{q}_\alpha{}^T\ \bar{\mathbf{q}}_\beta{}^T]\begin{Bmatrix}\bar{Q}_\alpha\\ \bar{Q}_\beta\end{Bmatrix}, \tag{2.20}$$

where $\mathbf{q}_\alpha$ is unknown while $\bar{\mathbf{q}}_\beta$, $\bar{\mathbf{Q}}_\alpha$, and $\bar{Q}_\beta$ are prescribed. Then the

* In the actual programming for digital computation the construction of these two assembled matrices is accomplished by efficient computer logic instructions instead of the matrix operations indicated here.

application of the variational principle $\delta\pi_p = 0$ will yield the system of equations*

$$\mathbf{K}_{\alpha\alpha}\mathbf{q}_\alpha = \bar{\mathbf{Q}}_\alpha - \mathbf{K}_{\alpha\beta}\,\bar{q}_\beta \tag{2.21}$$

which can be used to evaluate the unknown nodal displacements. Since the generalized displacements are the unknowns the finite element method is a matrix displacement method.

It is seen that the basic steps of the finite element methods outlined above are: (a) the determination of the element stiffness and force matrices based on an approximate solution of the elasticity problem for each individual element, (b) the construction of the final system of equations for the unknown parameters $\mathbf{q}$ using the matrices of the individual elements, and (c) the solution of the unknown nodal displacements. For the complete solution of the problem the displacement distribution over the volume must be determined using (2.7), and then the strain and stress distributions can be evaluated by (2.4) and (2.2) or (2.8) and (2.9).

The large majority of the existing finite element formulations are based on the assumed displacement approach. When the assumed displacements satisfy the completeness requirements, i.e. the representation of all rigid body displacements and the states of constant strain in the limit when the size of the element tends to zero and when the compatibility at the interelement boundaries are maintained, it is possible to prove the convergence of displacement solutions when the size of the element is progressively reduced (Key, 1966; Tong and Pian, 1967). The finite element formulation which is based on the compatible interelement displacement is named the compatible model.

For solutions of plane or three-dimensional elasticity problems for which only the first derivatives of the displacement functions appear in the variational functional only the continuity of the displacement components is required at the interelement boundary. In this case it is a relatively easy matter to construct interpolation functions to fulfill both completeness and compatibility conditions. The types of elements include triangular and quadrilateral elements for the plane problems and tetrahedron and hexa-

* The partition of the matrix in the form of (2.20) is not a practical procedure in actual programming. In practice, we obtained first from (2.18) a system of equations $\mathbf{Kq} = \mathbf{Q}$. When the ith generalized coordinate is prescribed to be $\bar{q}_i$, we first denote $\mathbf{K}_i$ as the ith column of $\mathbf{K}$, and replace $\bar{\mathbf{Q}}$ by $\bar{\mathbf{Q}} - \mathbf{K}_i\,\bar{q}_i$, except its ith component which is set to be $\bar{q}_i$. We then set the ith row and the ith column of $\mathbf{K}$ to be zero while the ith diagonal element is set to be unity. This procedure is performed for every constrained degree of freedom to obtain finally the constrained $\mathbf{K}'$ and $\bar{Q}'$. The equations to be solved are $\mathbf{K}'\mathbf{q} = \mathbf{Q}'$.

hedron elements for the three-dimensional solids. As discussed in the Appendix the interpolation functions may be linear or of high orders and there may be nodes located on the edges or faces of the elements in addition to those at the apexes of the elements. Furthermore, by the use of isoparametric transformation the interpolation functions for elements with curved sides or surfaces can be constructed.

For the plate and shell problems, however, when the Kirchhoff hypotheses are used, the second derivative of the lateral displacement w appears in the variational functional and the interelement compatibility condition would require the continuity of w as well as the normal slopes $w,_{\nu}$ at the interelement boundary. In this case the construction of the interpolation functions is no longer a simple task. Many compatible flat plate elements, however, have been constructed either by dividing the element into subregions, each using different interpolation functions, or by using derivatives of order higher than the first at the nodes. Indeed, the difficulty in constructing the compatible interpolation functions for plates and shells was the prime motivation for the development of other finite element models.

C. Illustrative Example of Compatible Model

To illustrate the procedure in the formulation of the compatible model a plane stress problem for a general orthotropic material is considered. As shown in Fig. 1, a uniform plate of thickness t is subjected to in-plane body forces $\bar{p}_x$ and $\bar{p}_y$ per unit volume and boundary tractions $\bar{T}_x$ and $\bar{T}_y$ per unit

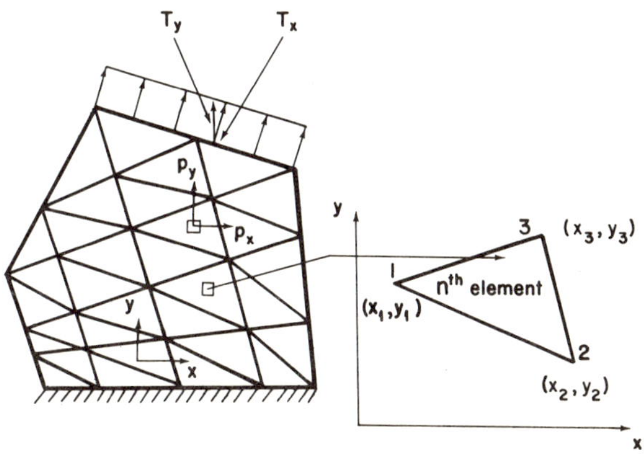

FIG. 1. Finite element modeling of a plane stress problem.

area respectively along the x and y directions. The plate is subdivided into triangular elements with the vertices of a typical one (nth element) located at $(x_1\, y_1)$, $(x_2\, y_2)$, and $(x_3\, y_3)$. The total potential energy π_p in this case is given by

$$\pi_p = t \sum_n \left(\int_{A_n} \tfrac{1}{2} \boldsymbol{\varepsilon}^T \mathbf{C} \boldsymbol{\varepsilon} \, dx\, dy - \int_{A_n} \mathbf{u}^T \bar{\mathbf{p}} \, dx\, dy - \int_{s_{\sigma n}} \mathbf{u}^T \bar{\mathbf{T}} \, ds \right), \tag{2.22}$$

where $\mathbf{u} = \{u\ v\}$ = displacement components,

$$\boldsymbol{\varepsilon} = \begin{Bmatrix} \varepsilon_{xx} \\ \varepsilon_{yy} \\ \varepsilon_{xy} \end{Bmatrix} = \begin{bmatrix} \partial/\partial x & 0 \\ 0 & \partial/\partial y \\ \partial/\partial y & \partial/\partial x \end{bmatrix} \begin{Bmatrix} u \\ v \end{Bmatrix} = \mathbf{D}\mathbf{u},$$

$$\mathbf{C} = \begin{bmatrix} C_{11} & C_{12} & C_{13} \\ C_{12} & C_{22} & C_{23} \\ C_{13} & C_{23} & C_{33} \end{bmatrix}, \tag{2.23}$$

$$\bar{\mathbf{p}} = \{\bar{p}_x\ \bar{p}_y\},$$
$$\bar{\mathbf{T}} = \{\bar{T}_x\ \bar{T}_y\}.$$

In the finite element formulation we assign the nodal displacements $\mathbf{q}_n$ for the nth element as

$$\mathbf{q}_n = \{q_1 \cdots q_6\} = \{u_1\ v_1\ u_2\ v_2\ u_3\ v_3\} \tag{2.24}$$

and interpolate u and v linearly by

$$\begin{Bmatrix} u \\ v \end{Bmatrix} = \begin{bmatrix} f_1 & 0 & f_2 & 0 & f_3 & 0 \\ 0 & f_1 & 0 & f_2 & 0 & f_3 \end{bmatrix} \begin{Bmatrix} q_1 \\ q_2 \\ \vdots \\ q_6 \end{Bmatrix} = \mathbf{f}\mathbf{q}, \tag{2.25}$$

where f_1, f_2, and f_3 are linear interpolation functions given in (A.14).

Here u and v are linear along each edge; hence if the nodal displacements coincide for two neighboring elements the compatibility of u and v along the interelement boundary is guaranteed. Substituting (2.23) and (2.24) into (2.22) we obtain, by taking $\mathbf{q}_n$ out of the integrals,

$$\pi_p = t \sum_n \left[\tfrac{1}{2} \mathbf{q}_n^T \int_{A_n} (\mathbf{D}\mathbf{f})^T \mathbf{C} (\mathbf{D}\mathbf{f}) \, dx\, dy \mathbf{q}_n - \mathbf{q}_n^T \left(\int_{A_n} \mathbf{f}^T \bar{\mathbf{p}} \, dx\, dy + \int_{s_{\sigma n}} \mathbf{f}^T \bar{\mathbf{T}} \, ds \right) \right] \tag{2.26}$$

By comparing this with Eq. (2.13) we obtain the following expressions for the element stiffness $\mathbf{k}_n$ and the generalized nodal forces $\bar{\mathbf{Q}}_n$;

$$\mathbf{k}_n = t \int_{A_n} (\mathbf{Df})^T \mathbf{C} (\mathbf{Df}) \, dx \, dy, \tag{2.27}$$

$$\bar{\mathbf{Q}}_n = t \left(\int_{A_n} \mathbf{f}^T \bar{\mathbf{p}} \, dx \, dy + \int_{s_{\sigma n}} \mathbf{f}^T \bar{\mathbf{T}} \, ds \right). \tag{2.28}$$

It is seen that the elements in the matrix $\mathbf{f}$ are all linear; hence the elements in $\mathbf{Df}$ are all constants. In fact,

$$\mathbf{Df} = \frac{1}{2A_n} \begin{bmatrix} y_{23} & 0 & y_{31} & 0 & y_{12} & 0 \\ 0 & x_{32} & 0 & x_{13} & 0 & x_{21} \\ x_{32} & y_{23} & x_{13} & y_{31} & x_{21} & y_{12} \end{bmatrix}, \tag{2.29}$$

where

$$x_{ij} = x_i - x_j; \; y_{ij} = y_i - y_j \tag{2.30}$$

and

$$\mathbf{k}_n = tA_n (\mathbf{Df})^T \mathbf{C} (\mathbf{Df}). \tag{2.31}$$

Equation (2.27) is the general form of the element stiffness matrix $\mathbf{k}$ derived by the compatible model. In general, the elements of that matrix are integrals over the discrete element. Since such integrals are usually evaluated by numerical quadratures (Irons, 1966) the matrix multiplications in (2.27) are performed by the computer, and to solve problems involving different elastic constants $\mathbf{C}$ would not need any additional algebraic manipulation.

D. Principle of Minimum Complementary Energy and Equilibrium Model I

For the finite element formulation the principle of minimum complementary energy may be stated as the vanishing of the variation of the following total complementary energy functional

$$\pi_c = \sum_n \left(\int_{V_n} \tfrac{1}{2} S_{ijkl} \sigma_{ij} \sigma_{kl} \, dV - \int_{S_{un}} T_i \bar{u}_i \, dS \right), \tag{2.32}$$

where S_{ijkl} is the elastic compliance tensor and S_{un} refers to portion of the boundary ∂V_n of the nth element over which the surface displacements u_i are prescribed. The complementary energy principle is also subjected to the condition of constraint that the stress equilibrium conditions (2.1) and

(2.5) are satisfied and the tractions at the interface of two neighboring elements are in equilibrium.

The conventional method in treating the stress equilibrium conditions in the theory of elasticity is the use of stress functions. Typical examples are the Airy stress functions to replace the inplane stresses of the plane stress and plane strain problems and the two stress functions U and V used by Fox and Southwell (1945) and many other authors (Fung, 1953; Morley, 1966) to replace the stress couples in the plate bending problems. Indeed, there exists the so-called static and geometric analogy between the stress function versus stress relations and the strain versus displacement relations for plate and shell problems (Southwell, 1950; Elias, 1967; Goldenveizer, 1940; Gunther, 1961; Reissner and Wan, 1969). For example, the formulation of the plate bending problem in terms of the stress functions U and V is analogous to the formulation of the plane stress (or plate stretching) problems in terms of the displacement components u and v. The former is based on the complementary energy principle while the latter is based on the potential energy principle.

It is apparent that there are also analogies between the corresponding finite element models (Fraeijs de Veubeke and Zienkiewicz, 1967; Elias, 1968). A finite element model which is based on the conventional complementary energy principle and uses stress functions as field variables is named equilibrium model I, because in this formulation the equilibrium conditions are satisfied everywhere. In the next subsection another equilibrium model will be discussed. The equilibrium model I for the plate bending problem is an exact analogy of the compatible model for the plane stress problems. Morley (1967, 1968) obtained solutions of plate bending problems using triangular equilibrium elements.

In the finite element formulation the stresses $\boldsymbol{\sigma}$ are first expressed in terms of stress functions U in the form

$$\boldsymbol{\sigma} = \mathbf{DU} + \bar{\boldsymbol{\Omega}}, \tag{2.33}$$

where $\mathbf{D}$ contains differential operators and $\bar{\boldsymbol{\Omega}}$ may be any particular solution of the equilibrium equation with the prescribed body forces, and hence are prescribed quantities. The boundary tractions are related to the stresses by (2.1.5). In certain applications such as plates and shells under Kirchhoff's hypothesis, the transverse shear along the boundary is related to the derivatives of the stress couples. The relation between $\mathbf{T}$ and $\boldsymbol{\sigma}$ is written in matrix form as

$$\mathbf{T} = \mathbf{N}\boldsymbol{\sigma}. \tag{2.34}$$

Analogous to the procedure described in Section II,B, the stress functions U are now approximated by interpolation functions over the individual elements in terms of the nodal generalized coordinates $\mathbf{p}_n$ which, for the present case, are the nodal values of the stress functions and may also be of their derivatives. We have

$$\mathbf{U} = \mathbf{B}\mathbf{p}_n, \tag{2.35}$$

where **B** is a matrix of interpolation functions. To maintain the equilibrium of the stresses at the interelement boundaries, the interpolation functions **B** must be so chosen that the stress functions U are continuous over the interelement boundaries.

Substituting (2.33), (2.34), and (2.35) into (2.32), we obtain

$$\pi_c = \sum_n^N (\tfrac{1}{2}\mathbf{p}_n{}^T \mathbf{l}_n \mathbf{p}_n - \mathbf{p}_n{}^T \bar{\mathbf{R}}_n + C_n), \tag{2.36}$$

where

$$\mathbf{l}_n = \int_{V_n} (\mathbf{DB})^T \mathbf{S} (\mathbf{D}B)\, dV,$$

$$\bar{\mathbf{R}}_n = \int_{V_n} \mathbf{S}\,\bar{\boldsymbol{\Omega}}\, dV - \int_{S_{u_n}} (\mathbf{DU})^T \mathbf{N}^T \bar{\mathbf{u}}\, dS, \tag{2.37}$$

$$C_n = \bar{\boldsymbol{\Omega}}^T \bar{\mathbf{u}} = \text{constant}.$$

In Eq. (2.37), **S** is the matrix of elastic compliance constants.

Equation (2.36) can be used to construct a system of equations with the nodal values of the stress functions **p** as the unknown parameters. When these quantities are determined, the distributions of the stress functions can be found by (2.33) and the stress and strain distributions are then obtained by the derivatives of the stress functions. The displacement distribution, however, can only be calculated by integrating the strain–displacement relation. In view of the fact that the stresses and strains provided by the finite element analysis are only approximate values, the displacements obtained are, in general, dependent upon the chosen integration path; hence they are not unique solutions. Another drawback is that when applying this equilibrium model it is often obscure to interpret the physical meaning of the boundary conditions for the stress functions.

E. Modified Complementary Energy Principle and Equilibrium Model II

In applying the complementary energy principle for finite element formulation it is obvious that the stress field need not be continuous across the interelement boundaries, while equilibrium conditions must be maintained for the boundary tractions T_i of two neighboring elements. If such equilibrium conditions are introduced as conditions of constraint and the corresponding Lagrangian multipliers—which can be interpreted as the displacements along the interelement boundary—are employed, the complementary functional may be rewritten as

$$\pi_{\mathrm{mc}} = \sum_n \left[\int_{V_n} \tfrac{1}{2} S_{ijkl}\sigma_{ij}\sigma_{kl}\, dV - \int_{S_n} T_i u_i \, dS - \int_{S_{u_n}} T_i \bar{u}_i \, dS \right], \tag{2.38}$$

where S_n is the portion of the boundary of ∂V_n which is connected with a neighboring element. It is seen that the present functional has the stresses within the elements and the displacements along the element boundaries as the field variables.

In the finite element formulation, the stresses $\boldsymbol{\sigma}$ may again be expressed in terms of (2.33); however, the stress functions $\mathbf{U}$ are not interpolated in terms of their nodal values but are of a series of independent functions with undetermined parameters $\boldsymbol{\beta}$. Such functions, for example, may be simply a polynomial expansion. Thus, the expressions for stresses $\boldsymbol{\sigma}$ are divided into two parts. The first part which consists of $\boldsymbol{\beta}$ must satisfy the homogeneous equations of equilibrium while the second part corresponds to a particular solution and is considered as prescribed. In matrix form it is convenient for later development to write

$$\boldsymbol{\sigma} = \mathbf{E}\boldsymbol{\beta} + \mathbf{E}_F \boldsymbol{\beta}_F, \tag{2.39}$$

where $\boldsymbol{\beta}$ is unknown and $\mathbf{E}_F\boldsymbol{\beta}_F$ is determined. For elements which contain boundaries with prescribed surface tractions, some of the β's in the first term are also prescribed. In that case, such prescribed β's are also put in the second term. The element boundary tractions are related to the assumed stresses by (2.34) and can be expressed as

$$\mathbf{T} = \mathbf{N}\mathbf{E}\boldsymbol{\beta} + \mathbf{N}\mathbf{E}_F\boldsymbol{\beta}_F. \tag{2.40}$$

In order to maintain the equilibrium along the interelement boundary the tractions $\mathbf{T}$ along the common boundary of each pair of neighboring elements will be represented by a number of generalized forces $\mathbf{R}$ where

$$\mathbf{T} = \boldsymbol{\phi}\mathbf{R}. \tag{2.41}$$

By defining $\mathbf{G}$ and $\mathbf{G}_F$ by

$$\mathbf{N}\mathbf{E} = \boldsymbol{\phi}\mathbf{G}^T \tag{2.42}$$

and

$$\mathbf{N}\mathbf{E}_F = \boldsymbol{\phi}\mathbf{G}_F^{\,T},$$

we obtain

$$\mathbf{T} = \boldsymbol{\phi}\mathbf{G}^T\boldsymbol{\beta} + \boldsymbol{\phi}\mathbf{G}_F^{\,T}\boldsymbol{\beta}_F. \tag{2.43}$$

We define the generalized boundary displacements $\mathbf{q}_n$ by

$$\mathbf{R}^T\mathbf{q}_n = \int_{\partial V_n} T_i u_i \, dS = \mathbf{R}^T \int_{\partial V_n} \boldsymbol{\phi}^T \mathbf{u}_B \, dS, \tag{2.44}$$

where $\mathbf{u}_B$ are the displacements along the boundary. Thus the generalized displacements are

$$\mathbf{q}_n = \int_{\partial V_n} \boldsymbol{\phi}^T \mathbf{u}_B \, dS \tag{2.45}$$

and can be interpreted as the weighted averages of the boundary displacements.

Substituting (2.39) and (2.43) into (2.38) and using (2.45) we obtain

$$\pi_{\mathrm{mc}} = \sum_n (\tfrac{1}{2}\boldsymbol{\beta}^T\mathbf{H}\boldsymbol{\beta} + \boldsymbol{\beta}^T\mathbf{H}_F\boldsymbol{\beta}_F - \boldsymbol{\beta}^T\mathbf{G}\mathbf{q}_n + \mathbf{Z}^T\mathbf{q}_n + B_n), \tag{2.46}$$

where

$$\begin{aligned} \mathbf{H} &= \int_{V_n} \mathbf{E}^T\mathbf{S}\mathbf{E} \, dV, \\ \mathbf{H}_F &= \int_{V_n} \mathbf{E}^T\mathbf{S}\mathbf{E}_F \, dV, \\ \mathbf{Z}^T &= -\boldsymbol{\beta}_F{}^T\mathbf{G}_F, \\ B_n &= \tfrac{1}{2}\boldsymbol{\beta}_F{}^T \int_{V_n} \mathbf{E}_F{}^T\mathbf{S}\mathbf{E}_F \, dV \boldsymbol{\beta}_F = \text{constant}. \end{aligned} \tag{2.47}$$

The stationary condition of the functional given by (2.46) with respect to variations of $\boldsymbol{\beta}$ then yields

$$\mathbf{H}\boldsymbol{\beta} + \mathbf{H}_F\boldsymbol{\beta} - \mathbf{G}\mathbf{q}_n = 0 \tag{2.48}$$

for each individual elements. By solving for $\boldsymbol{\beta}$ in terms of $\mathbf{q}_n$ and substituting back into (2.46) we can express the functional π_{mc} in terms of $\mathbf{q}_n$ only, i.e.

$$\pi_{\mathrm{mc}} = -\sum_n (\tfrac{1}{2}\mathbf{q}_n{}^T\mathbf{k}_n \mathbf{q}_n - \mathbf{q}_n{}^T \bar{\mathbf{Q}}_n + C_n), \tag{2.49}$$

where

$$\begin{aligned} \mathbf{k}_n &= \mathbf{G}^T\mathbf{H}^{-1}\mathbf{G}, \\ \bar{\mathbf{Q}}_n &= \mathbf{G}^T\mathbf{H}^{-1}\mathbf{H}_F\boldsymbol{\beta}_F + \mathbf{Z}, \end{aligned} \tag{2.50}$$

and C_n is a constant.

Upon comparing (2.49) with (2.13), it is clear that, here again, $\mathbf{k}_n$ and $\bar{\mathbf{Q}}_n$ are respectively the element stiffness matrix and the vector of generalized forces. It is also clear that further development of this model will be identical to that in Section II,B and the resulting finite element method is also a matrix displacement method.

This equilibrium model, which is now classified as equilibrium model II, was suggested by Fraeijs de Veubeke (1964) and has been used to analyze numerous plate bending problems by his research group (Fraeijs de Veubeke, 1965, 1966; Fraeijs de Veubeke and Sander, 1968). It is clear that the difficulty for constructing the fully compatible interpolation functions no longer appears in the present equilibrium model.

F. Modified Complementary Energy Principle and the Hybrid Stress Model

The modified complementary energy principle may be used to derive another finite element model. In this formulation the construction of the assumed stresses $\boldsymbol{\sigma}$ within each element follows the same procedure as the previous subsection; however, instead of the consideration of equilibrium of boundary tractions $\mathbf{T}$, the boundary displacements $\mathbf{u}_B$ are now interpolated in terms of a finite number of generalized boundary displacements $\mathbf{q}_n$,

$$\mathbf{u}_B = \mathbf{L}\mathbf{q}_n . \tag{2.51}$$

Here the interpolation functions $\mathbf{L}$ are applied only to individual boundary segments. It is still necessary to construct these interpolation functions so that when the nodal displacements of two neighboring elements coincide, the displacements along the entire common boundary are compatible. The present model is called a hybrid stress model to reflect the fact that the assumed stresses are in equilibrium only within each element, but the compatibility of the displacements along the interelement boundaries is considered.

In formulating this model (2.38) may be followed. In such a case the assumed stresses $\boldsymbol{\sigma}$ must satisfy the prescribed boundary tractions along $S_{\sigma n}$ and the boundary displacement interpolation (2.51) is no longer needed for this portion of the boundary. Thus the elements which involve prescribed tractions and those which do not have prescribed tractions will have to be treated differently. A simpler finite element formulation, however, can be accomplished by using a slightly modified version of (2.38). When the conditions $T_i - \bar{T}_i = 0$ along $S_{\sigma n}$ are introduced as conditions of constraint, with the corresponding boundary displacements u_i as the Lagrangian multipliers, (2.38) can be rewritten as

$$\pi_{\mathrm{mc}} = \sum_n \left[\int_{\partial V_n} \tfrac{1}{2} S_{ijkl} \sigma_{ij} \sigma_{kl} \, dV - \int_{\partial V_n} T_i u_i \, dS + \int_{S_{\sigma n}} \bar{T}_i u_i \, dS \right]. \quad (2.52)$$

The displacements u_i now appear over the entire boundary ∂V_n of all elements; hence in the finite element formulation the boundary displacement interpolation (2.51) is applied to all element boundaries. Substituting (2.39), (2.40), and (2.51) into either (2.38) or (2.52) will yield an expression for π_{mc} which is identical to (2.46) except that the matrices $\mathbf{G}$ and $\mathbf{G}_F$ are now defined by

$$\begin{aligned} \mathbf{G} &= \int (\mathbf{NE})^T \mathbf{L} \, dS, \\ \mathbf{G}_F &= \int (\mathbf{NE}_F)^T \mathbf{L} \, dS. \end{aligned} \quad (2.53)$$

The matrix $\mathbf{Z}^T$ is the same as that in (2.47) if the assumed stresses $\boldsymbol{\sigma}$ satisfy the prescribed tractions along $S_{\sigma n}$; hence (2.38) is used. Otherwise, if (2.52) is used,

$$\mathbf{Z}^T = -\boldsymbol{\beta}_F \mathbf{G}_F + \int_{S_{\sigma n}} \bar{\mathbf{T}} \mathbf{L} \, dS. \quad (2.54)$$

From this step on the development of the finite element hybrid stress model is identical to that of equilibrium model II. Obviously, the resulting finite element method of the hybrid stress model is also a matrix displacement method.

For both the compatible model and the hybrid stress model the element stiffness matrices $\mathbf{k}_n$ are based on the element nodal displacements $\mathbf{q}_n$. We have mentioned that in formulating a compatible model difficulties may often occur in constructing compatible interpolation functions for displacements over the entire element as well as along the interelement boundary. For the formulation of the hybrid stress model it is only necessary to interpolate the displacements individually along portions of the element boundary; hence it is a much easier task to construct the appropriate compatible functions.

It should be noted that when the displacements u_i are fixed on the interelement boundaries π_{mc} is a minimum with respect to the stress variables σ_{ij}. However, π_{mc} becomes a maximum with respect to u_i when σ_{ij} are expressed in terms of u_i. In the finite element solution, the increase of the number of the stress parameters β will tend to decrease the value of π_{mc}, while the increase of the number of the boundary displacement parameters will tend to increase π_{mc}. Thus there exists an optimum relation between the number of β's and q's. In practice this optimum is not known. However, it is advisable to choose the stress variables and the displacement variable so that they maintain the same order of approximation (Pian and Tong, 1969a).

By examining the expression for the element stiffness matrix $\mathbf{k}_n$ given by (2.50), (2.47), and (2.53), it is easy to verify that the derivation of this matrix involves essentially an approximate solution for an element with prescribed boundary displacements which are interpolated in terms of a finite number of nodal displacements and are compatible with those of the neighboring elements. Previously, for the compatible model, such a problem was tackled by interpolating the displacements within the element. For the present hybrid model the problem is solved by assuming stress distributions and hence by the use of the principle of minimum complementary energy. This idea for deriving the element stiffness matrix, in fact, was used initially (Pian, 1964, 1966) prior to the establishment of the modified complementary principle and the formal derivation of the hybrid stress model (Tong and Pian, 1969).

The hybrid stress model is especially convenient for treating problems which involve stress singularities. For example, for the elastic solution of a plane stress panel involving a sharp crack the behavior of the singular stress term at the crack tip is known while its magnitude, which is generally called the stress intensity factor in the nomenclature of fracture mechanics, is an unknown parameter to be determined. In applying the hybrid stress model it is only necessary to include such a term in the assumed stress functions for the group of elements at the crack tip. The nonsingular stress parameter $\boldsymbol{\beta}$ in each element may be eliminated as usual while the stress intensity factor is left as an unknown in addition to the nodal displacement $\mathbf{q}$. The hybrid stress model has been found to be a very efficient method for estimating the stress intensity factors (Pian *et al.*, 1972).

G. Illustrative Example of Hybrid Stress Model

The procedure in the formulation of the hybrid stress model is now illustrated by the derivation of the element stiffness matrix of a multilateral flat plate element in bending. Under the Kirchhoff hypothesis the complementary strain energy in the plate is the resultant of the stress couples $M_{\alpha\beta}$ ($\alpha, \beta = 1, 2$) only and the functional π_{mc} in (2.52) for a plate which is subdivided into a finite number of elements may be expressed as

$$\pi_{\mathrm{mc}} = \sum_n \left[\int_{A_n} \tfrac{1}{2} D_{\alpha\beta\gamma\delta} M_{\alpha\beta} M_{\gamma\delta}\, dA - \int_{\partial A_n} (Qw + m_\alpha w_{,\alpha})\, ds \right.$$

$$\left. + \int_{S_{\sigma n}} (\bar{Q}w + \bar{m}_\alpha w_{,\alpha})\, ds \right] \qquad (2.55)$$

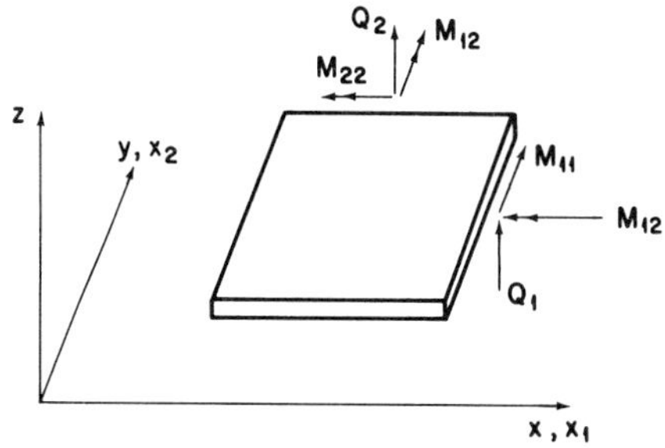

FIG. 2. Sign convention for stress resultants and couples in a plate.

where

$M_{\alpha\beta}$ = stress couples (with sign convention indicated in Fig. 2),
$D_{\alpha\beta\gamma\delta}$ = flexural rigidity constants relating the stress couples and the plate curvatures,
A_n = area of the nth element,
∂A_n = boundary of the nth element,
Q = transverse shear force per unit length of the element boundary,
m_α = components of resulting moment per unit length of the element boundary,
$s_{\sigma n}$ = portion of the element boundary where the tractions are prescribed.

In the above equation repeated Greek indices imply summation over 1 and 2.

The stress couples $M_{\alpha\beta}$, the transverse shear stress resultants Q_α, and the intensity of the distributed lateral loading p (per unit area) are related by the following equilibrium equations:

$$M_{\alpha\beta,\beta} = Q_\alpha$$

and (2.56)

$$Q_{\alpha,\alpha} + p = 0,$$

which may be combined into one equation for $M_{\alpha\beta}$, i.e.

$$M_{\alpha\beta,\alpha\beta} + p = 0. \tag{2.57}$$

The shear force Q and the components of the resulting moment m_α ($\alpha = 1, 2$) along the boundary are related to the stress couples $M_{\alpha\beta}$ and stress resultants Q_α by

$$m_\alpha = -M_{\alpha\beta}\nu_\beta$$

and (2.58)

$$Q = Q_\alpha \nu_\alpha,$$

and

$$
\begin{aligned}
H_{01}^{(1)}(s) &= 1 - 3\xi^2 + 2\xi^3, \\
H_{02}^{(1)}(s) &= 3\xi^2 - 2\xi^3, \\
H_{11}^{(1)}(s) &= l[\xi - 2\xi^2 + \xi^3], \\
H_{12}^{(1)}(s) &= l[\xi^3 - \xi^2],
\end{aligned} \tag{2.68}
$$

where $\xi = s/l$ and l is the length of the edge. The derivative $w_{,s}$ can be obtained from (2.65), i.e.

$$
w_{,s}(s) = \frac{dH_{01}^{(1)}}{ds} w_1 + \frac{dH_{02}^{(1)}}{ds} w_2 + \frac{dH_{11}^{(1)}}{ds} (w_{,s})_1 + \frac{dH_{12}^{(1)}}{ds} (w_{,s})_2 . \tag{2.69}
$$

In order to express $w(s)$ and $w_{,\alpha}(s)$ in terms of the nodal values of w and $w_{,\alpha}$, it is only necessary to use (2.67) and to express $w_{,\alpha}(s)$ by

$$
\begin{aligned}
w_{,1}(s) &= -\nu_2 w_{,s}(s) + \nu_1 w_{,\nu}(s), \\
w_{,2}(s) &= \nu_1 w_{,s}(s) + \nu_2 w_{,\nu}(s),
\end{aligned} \tag{2.70}
$$

where $w_{,s}(s)$ and $w_{,\nu}(s)$ are given by (2.69) and (2.66), and then to substitute the nodal values of $w_{,s}$ and $w_{,\nu}$ by

$$
\begin{aligned}
w_{,\nu} &= w_{,1}\nu_1 + w_{,2}\nu_2 , \\
w_{,s} &= -w_{,1}\nu_2 + w_{,2}\nu_1 .
\end{aligned} \tag{2.71}
$$

Thus for a typical edge m the boundary displacement matrix is

$$
(\mathbf{u}_B)_m = \{w_m (w_{,1})_m (w_{,2})_m\} = \mathbf{L}_m \mathbf{q}, \tag{2.72}
$$

where $\mathbf{L}_m$ is defined as follows: Let

$$
\begin{aligned}
\mathbf{L}_{1m} &= \begin{bmatrix} 1 & 0 & 0 \\ 0 & -\nu_2 & \nu_1 \\ 0 & \nu_1 & \nu_2 \end{bmatrix}, \\
\mathbf{L}_{2m} &= \begin{bmatrix} H_{01}^{(1)} & H_{11}^{(1)} & 0 \\ \dfrac{dH_{01}^{(1)}}{ds} & \dfrac{dH_{11}^{(1)}}{ds} & 0 \\ 0 & 0 & H_{01}^{(1)} \end{bmatrix}, \\
\mathbf{L}_{3m} &= \begin{bmatrix} H_{02}^{(1)} & H_{12}^{(1)} & 0 \\ \dfrac{dH_{02}^{(1)}}{ds} & \dfrac{dH_{12}^{(1)}}{ds} & 0 \\ 0 & 0 & H_{02}^{(0)} \end{bmatrix},
\end{aligned} \tag{2.73}
$$

and

$$
\begin{aligned}
\mathbf{L}_m' &= \mathbf{L}_{1m} \mathbf{L}_{2m} \mathbf{L}_{1m} , \\
\mathbf{L}_m'' &= \mathbf{L}_{1m} \mathbf{L}_{3m} \mathbf{L}_{1m} .
\end{aligned} \tag{2.74}
$$

The matrix $\mathbf{L}_m$ can be expressed in terms of $\mathbf{L}_m'$ and $\mathbf{L}_m''$. Their relations will depend on the particular side m of the element. We use a quadrilateral element (Fig. 3) as an example. The four sides are labeled a, b, c, and d, and the four corners, A, B, C, and D. Let the column matrix $\mathbf{q}$ be arranged in the following manner,

$$\begin{aligned}\mathbf{q} = \{&w_A(w_{,1})_A(w_{,2})_A\, w_B(w_{,1})_B(w_{,2})_B \\ &w_C(w_{,1})_C(w_{,2})_C\, w_D(w_{,1})_D(w_{,2})\}_D.\end{aligned} \tag{2.75}$$

Then

$$\begin{aligned}\mathbf{L}_m = &\{\mathbf{L}_{a'}\ \mathbf{L}_{a''}\ 0\quad 0\} &&\text{on side } a,\\ &\{0\quad \mathbf{L}_{b'}\ \mathbf{L}_{b''}\ 0\} &&\text{on side } b,\\ &\{0\quad 0\quad \mathbf{L}_{c'}\ \mathbf{L}_{c''}\} &&\text{on side } c,\\ &\{\mathbf{L}_{d'}\ 0\quad 0\quad \mathbf{L}_{d''}\} &&\text{on side } d.\end{aligned} \tag{2.76}$$

Finally, the flexural rigidity constants which relate the stress couples and the curvatures may be written in matrix form,

$$\mathbf{D} = \begin{bmatrix} D_{1111} & D_{1122} & 2D_{1112} \\ D_{1122} & D_{2222} & 2D_{2212} \\ 2D_{1112} & 2D_{2212} & 4D_{1212} \end{bmatrix}. \tag{2.77}$$

Given $\mathbf{E}$ by (2.60), $\mathbf{N}_m$ by (2.64), and $\mathbf{L}_m$ by (2.73) through (2.76), the element stiffness matrix can be obtained from

$$\mathbf{k}_n = \mathbf{G}^T\mathbf{H}^{-1}\mathbf{G}, \tag{2.78}$$

where

$$\begin{aligned}\mathbf{H} &= \int_{A_n} \mathbf{E}^T\mathbf{D}\mathbf{E}\, dA,\\ \mathbf{G} &= \int_{\partial A_n} (\mathbf{N}\mathbf{E})^T\mathbf{L}\, ds = \sum_m \int_m (\mathbf{N}_m\mathbf{E})^T\mathbf{L}_m\, ds,\end{aligned} \tag{2.79}$$

where $\sum_m$ denotes summation over all sides of the element, i.e. the integrations along the boundary are performed separately for each side.

H. Modified Potential Energy Principles and Hybrid Displacement Models

In applying the conventional principle of minimum potential energy the functional to be varied involves only the displacement field which is to be continuous over the entire solid continuum. In the finite element application, however, it is permissible to modify the functional by introducing the displacement compatibility conditions along the interelement boundaries as conditions of constraints and hence to include the corresponding Lagrangian multipliers, the interelement boundary tractions T_i, as additional variables. The modified variational functional becomes

$$\pi_{\mathrm{mp1}} = \sum_n \Bigg[\int_{V_n} (\tfrac{1}{2} C_{ijkl} \varepsilon_{ij} \varepsilon_{kl} - \bar{F}_i u_i) \, dV - \int_{S_n} T_i u_i \, dS - \int_{S_{un}} T_i (u_i - \bar{u}_i) \, dS - \int_{S_{\sigma n}} \bar{T}_i u_i \, dS \Bigg]. \tag{2.80}$$

Here the strain components ε_{ij} are again expressed in terms of the displacements. This variational principle thus has the displacement field and the element boundary tractions as variables. The use of this principle was first suggested by Jones (1964) and has been used by Yamamoto (1966) and Greene *et al.* (1969) for finite element formulations.

In formulating the finite element model using this variational principle, the element displacements **u** are approximated by a finite number of terms with unknown parameters $\boldsymbol{\alpha}$, and the boundary tractions **T** are interpolated in terms of generalized internal forces **R**. This model is named a hybrid displacement model to indicate that the assumed displacement are continuous only within each element while the interelement forces are in equilibrium. The expression for π_{mp1} is thus in terms of $\boldsymbol{\alpha}$ and **R**. By taking the variation of π_{mp1} with respect to $\boldsymbol{\alpha}$ and **R** a system of equations can be obtained with $\boldsymbol{\alpha}$ and **R** as unknowns.

One might think that since the hybrid stress model yields a matrix displacement method the present hybrid displacement model should yield correspondingly a matrix force method if the parameters $\boldsymbol{\alpha}$ are eliminated first. It turns out that the resulting matrix equations do not correspond to a matrix force method which should contain only a set of redundant force **X** as unknowns. It is clear that not all the internal forces **R** are redundant forces. In general, in the formulation of this hybrid model both $\boldsymbol{\alpha}$ and **R** are left as the unknowns and the corresponding finite element method is a matrix mixed method.

The approximation of **u** within each element may be simple polynomials such as the solutions by Yamamoto (1966) and Greene *et al.* (1969). Another way of applying this modified principle is to interpolate the element displacements **u** in terms of nodal displacements **q**, to guarantee the continuity with the neighboring element at the nodes but not the complete compatibility conditions along the interelement boundaries. Such conditions are then introduced by Lagrangian multipliers which are the generalized internal forces **R**. The final expression for π_{mp1} is thus in terms of both

$\mathbf{q}$ and $\mathbf{R}$, hence the corresponding finite element method is again a matrix mixed method. Harvey and Kelsey (1971) applied this technique in solving plate bending problems using triangular elements. For each element a complete cubic polynomial is used to express the lateral displacement w in terms of generalized displacements w, $w_{,1}$, and $w_{,2}$ at the corner nodes. This interpolation will satisfy the continuity of w and $w_{,s}$ at the interelement boundaries but since the normal slope $w_{,\nu}$ varies quadratically along the edge its continuity across the boundary cannot be satisfied if the conventional assumed displacement method is employed. By using the modified potential energy principle, conditions of constraint can be applied to enforce the continuity of the normal slopes along the interelement boundary. In the present case, such a condition is merely the equating of the normal slopes of two neighboring elements at the center of the edge. The corresponding Lagrangian multipliers $\boldsymbol{\lambda}$, of course, are in the unit of bending moment. The resulting matrix equations obtained by Harvey and Kelsey are in the form of

$$\begin{bmatrix} \mathbf{K} & \mathbf{B}^T \\ \mathbf{B} & \mathbf{0} \end{bmatrix} \begin{bmatrix} \mathbf{q} \\ \boldsymbol{\lambda} \end{bmatrix} = \begin{bmatrix} \mathbf{Q} \\ \mathbf{0} \end{bmatrix}, \tag{2.81}$$

where $\mathbf{0}$ is either a diagonal matrix or a column matrix of zeros.

Another modification of the potential energy principle is due to Tong (1970) and is based on the use of separate variables for the interior displacement field and the interelement boundary displacements. The compatibility condition of these two variables at the element boundary can be introduced by means of Lagrangian multipliers, which can again be recognized as the boundary tractions but are, in this case, independent for the two neighboring elements. The functional to be varied under this principle is

$$\begin{aligned} \pi_{\mathrm{mp2}} = \sum_n \bigg[\int_{V_n} (\tfrac{1}{2} C_{ijkl} \varepsilon_{ij} \varepsilon_{kl} - \bar{F}_i u_i)\, dV - \int_{\partial V_n} T_i (u_i - \tilde{u}_i)\, dS \\ - \int_{S_{\sigma n}} \bar{T}_i u_i \, dS - \int_{S_{un}} T_i (\bar{u}_i - \tilde{u}_i)\, dS \bigg], \end{aligned} \tag{2.82}$$

where $\tilde{u}_i$ is the interelement boundary displacement. The variables in this variational principle are the element displacements u_i, the interelement boundary displacements $\tilde{u}_i$, and the element boundary tractions T_i.

In applying the finite element formulation the element displacements $\mathbf{u}$ are again approximated in terms of a finite number of parameters $\boldsymbol{\alpha}$, while the boundary tractions $\mathbf{T}$ and the boundary displacements $\tilde{\mathbf{u}}$ are interpolated in terms of generalized internal forces $\mathbf{R}$ and nodal displacements $\mathbf{q}$, respectively. Here $\boldsymbol{\alpha}$ and $\mathbf{R}$ for one element are all independent of those for other elements; hence they can be solved in terms of the generalized nodal displacements $\mathbf{q}$. The resulting expression for π_{mp2} is of the same form as π_{mc} in (2.49) and the resulting finite element method is a matrix displacement method. The hybrid displacement model II by Tong was used to solve plate bending problems (Tong, 1970) and shell problems (Atluri, 1969).

I. Reissner's Variational Principle and the Mixed Model

Reissner's variational principle (Reissner, 1950) states that a functional π_{R} is stationary with respect to the variation of both the displacements u_i and stresses σ_{ij} where

$$\begin{aligned}\pi_{\mathrm{R}} = \int_V [-\tfrac{1}{2}S_{ijkl}\sigma_{ij}\sigma_{kl} + \tfrac{1}{2}\sigma_{ij}(u_{i,j} + u_{j,i}) \\ - \bar{F}_i u_i]\, dV - \int_{S_\sigma} \bar{T}_i u_i \, dS - \int_{S_u} T_i(u_i - \bar{u}_i)\, dS.\end{aligned} \tag{2.83}$$

In the finite element analysis, stresses and displacements are assumed separately for each individual element. Unlike the complementary energy principle, the assumed stresses here are not required to satisfy the equilibrium equations. The continuity requirement for the stresses and displacements at the interelement boundaries are also relaxed so that the combination of these quantities need only satisfy the condition that the functional π_{R} is defined. For example, in the case of a plane stress problem if a local coordinate system ν–s is used to denote the normal and tangential directions at the boundary of an element, then the second term of the first integral can be written as

$$I = \int_A [\sigma_{\nu\nu} u_{\nu,\nu} + \sigma_{\nu s}(u_{\nu,s} + u_{s,\nu}) + \sigma_{ss} u_{s,s}]\, dA. \tag{2.84}$$

It is seen that this integral is still defined even if the displacements are discontinuous across the boundary so long as the stresses are continuous. Also, if the displacements are continuous the corresponding stresses will not be necessarily continuous in order to make the above integral finite. It is understood that in the finite element formulation the displacements and stresses within each element are continuous along the boundary. Therefore, for the plane stress problem there are four possible combinations for the compatibility requirement across the interelement boundary:

a. Both stress components $\sigma_{\nu\nu}$ and $\sigma_{\nu s}$ are continuous.
b. Normal stress $\sigma_{\nu\nu}$ and tangential displacement u_s are continuous.
c. Normal displacement u_ν and shear stress $\sigma_{\nu s}$ are continuous. Or,
d. Both displacement components u_ν and u_s are continuous.

In the case of plate bending the energy functional π_R takes the form

$$\begin{aligned}\pi_R = \int_A [-\tfrac{1}{2}D_{\alpha\beta\gamma\delta} M_{\alpha\beta} M_{\gamma\delta} - M_{\alpha\beta} w_{,\alpha\beta} - pw]\, dA \\ - \int_{S\sigma} (\bar{m}_\nu w_{,\nu} + \bar{m}_s w_{,s} + \bar{Q}w)\, ds \\ - \int_{Su} [m_\nu(w_{,\nu} - \bar{w}_{,\nu}) + m_s(w_{,s} - \bar{w}_{,s}) + Q(w - \bar{w})]\, ds,\end{aligned} \tag{2.85}$$

where m_ν and m_s are the components of moment over unit length of the boundary, the Greek indices α, β, γ, and δ take the value of 1 or 2, and repeated indices refer to summation over 1 and 2. Again by referring to a local coordinate system ν–s, the second term of the first integral can be written as

$$I = \int_A (M_{\nu\nu} w_{,\nu\nu} + 2M_{\nu s} w_{,\nu s} + M_{ss} w_{,ss})\, dA. \tag{2.86}$$

This integral is definable if any of the following sets of conditions is satisfied across the interelement boundary:

a. continuity of w and $w_{,\nu}$,
b. continuity of w and m_ν,
c. continuity of m_ν and $m_{\nu,\nu}$.

For the finite element formulation (2.85) may be written as

$$\begin{aligned}\pi_R = \sum_n \Big\{ \int_{An} (-\tfrac{1}{2}D_{\alpha\beta\gamma\delta} M_{\alpha\beta} M_{\gamma\delta} - M_{\alpha\beta} w_{,\alpha\beta} - pw)\, dA \\ + I_n - \int_{S\sigma n} (\bar{m}_\nu w_{,\nu} + \bar{m}_s w_{,s} + \bar{Q}w)\, ds \\ - \int_{Sun} [m_\nu(w_{,\nu} - \bar{w}_{,\nu}) + m_s(w_{,s} - \bar{w}_{,s}) + Q(w - \bar{w})]\, ds \Big\},\end{aligned} \tag{2.87}$$

where I_n are defined differently under the following continuity conditions across the interelement boundary:

a. When w and $w_{,\nu}$ are continuous

$$I_n = 0. \tag{2.88}$$

b. When w and m_ν are continuous but $w_{,\nu}$ are discontinuous

$$I_n = -\int_{sn} m_\nu w_{,\nu}\, ds, \tag{2.89}$$

where s_n is the portion of the element boundary over which a discontinuity in $w_{,\nu}$ exists.

c. When w, $w_{,\nu}$, and $w_{,s}$ are discontinuous while m_s, m_ν, and the normal derivative of m_ν are continuous

$$I_n = -\int_{s_n} [m_\nu w_{,\nu} + m_s w_{,s} + w(m_{\nu,\nu} + m_{s,s})]\, ds. \tag{2.90}$$

In the finite element formulation by Reissner's variation principle both stresses and displacements are assumed within each element; thus the finite element model is classified as a mixed model. The finite element formulations for the different interelement continuity conditions may be quite different. For example, in applying (2.87) when w and $w_{,\nu}$ are continuous then the stresses $\boldsymbol{\sigma}$ may be completely independent for different elements. In such a case $\boldsymbol{\sigma}$ for each element may be expressed in terms of a finite number of parameters $\boldsymbol{\beta}$, and the displacements $\mathbf{u}$ interpolated in terms of generalized nodal displacements. It is obvious that such a formulation will lead to an expression for π_R similar to (2.46) which is associated with the equilibrium model II and the hybrid stress model. Thus, the resulting matrix equations will contain only the nodal displacements as the generalized coordinates. The application of the Reissner's principle in this form, however, still requires the construction of interpolation functions over the elements such that the interelement compatibility conditions are fulfilled. Under such conditions, the original compatible model would be a simpler choice. Similarly in applying (2.87) when w and $w_{,\nu}$ are discontinuous the final matrix equations may have only the nodal values of the stresses as unknowns.

A distinct form of the finite element equations can be obtained if the Reissner's variational principle is used which requires partial fulfillment of the continuity of both displacements and stresses. For example, for the plate bending problem given by (2.87) if only the continuity of w and m_ν is required across the interelement boundaries the interpolation functions for both displacement w and stress couples $M_{\alpha\beta}$ can be easily constructed.

In applying the Reissner's variational principle for finite element analysis of the plate bending problem, Herrmann (1965, 1967) employed the condition (b) above but used a different version of the energy functional. When (2.87) is integrated by parts we obtain

$$\begin{aligned}\pi_R = \sum_n \Big\{ \iint_{A_n} [-\tfrac{1}{2} D_{\alpha\beta\gamma\delta} M_{\alpha\beta} M_{\gamma\delta} + w_{,\alpha} M_{\alpha\beta,\beta} - pw]\, dA + \int_{s_n} m_s w_{,s}\, ds \\ - \int_{s_{\sigma n}} [(\bar{m}_\nu - m_\nu) w_{,\nu} + (\bar{m}_s - m_s) w_{,s} + \bar{Q} w]\, ds \\ + \int_{s_{un}} [m_\nu \bar{w}_{,\nu} + m_s \bar{w}_{,s} - Q(w - \bar{w})]\, dt \Big\}.\end{aligned} \tag{2.91}$$

The second term in the above integral is introduced to account for a possible discontinuity in m_s across the interelement boundary. Herrmann (1967), in fact, formulated his plate bending solution by using triangular elements in each of which the lateral displacement was assumed linear while the stress couples were assumed constants. The generalized coordinates for each element are the three nodal displacements $\mathbf{q}$ and the normal moments m_ν along the three edges. In an earlier paper (Herrmann, 1966) he assumed that all the stress couples are linear within each element and are continuous along the interelement boundary; hence he needed to use nine generalized coordinates for stress couples for a triangular element. In this case the continuity condition is more than it is necessary.

In the finite element formulation by Reissner's Principle the stresses $\boldsymbol{\sigma}$ are expressed in terms of generalized boundary tractions $\mathbf{R}$, and the displacements $\mathbf{u}$ are interpolated in terms of a generalized nodal displacement $\mathbf{q}$. The resulting equations of this formulation, in general, contain both $\mathbf{q}$ and $\mathbf{R}$ as unknowns. The finite element method is thus a matrix mixed method.

J. Remarks

Of all the finite element models presented in the previous subsections, the conventional assumed displacement compatible model is still, by far, the simplest scheme if an appropriate interpolation function can be constructed that will also satisfy the interelement compatibility conditions. For problems such as plane elasticity, axisymmetric solids, and three-dimensional solids for which the continuity of the normal derivatives along the interelement boundaries is not required it is an easy task to formulate the finite element method by the compatible model. The finite element models, other than the compatible model, therefore have been applied almost exclusively to plate and shell problems for which it is difficult to construct a completely compatible interpolation function. One exception, perhaps, is the application of the finite element method to problems involving stress singularities for which a hybrid model can be easily adapted to include the singularities (Pian *et al.*, 1972).

Historically, in the development of the finite element analyses, the matrix displacement method was adopted because its resulting symmetric positive definite matrix is, in general, well conditioned and can be easily handled by the digital computers. Thus, the existing general purpose computer programs for the analysis of solid continua are based almost inclusively on the matrix displacement scheme. In order to use these programs the finite element model to be selected must lead to a matrix displacement method. It is fortunate that in addition to the compatible model, the equilibrium model II, the hybrid stress model, and the hybrid displacement model II all lead to the matrix displacement scheme.

The finite element solutions by the equilibrium model II and the compatible model are based on minimum principles, and hence can be proved (Fraeijs de Veubeke, 1964, 1965a) to yield respectively the upper and lower bounds of the strain energy stored in the solid for problems with prescribed external loads. Thus, when finite element solutions can be obtained by using both models, the accuracy of the numerical solutions can be assessed. The hybrid stress model is based only on a stationary principle; hence its solution may yield a strain energy either higher or lower than the exact value. The strain energy obtained by the hybrid stress model, however, can be proved to be bounded always from above by an equilibrium model solution provided they have the same type of stress distribution within each element; and always from below by a compatible model solution if they have the same type of displacement distribution along the interelement boundary (Tong and Pian, 1970).

A comparison is made of the finite element solutions of a centrally loaded square plate with clamped boundary analyzed by the three models, the equilibrium model II, the compatible model, and the hybrid stress models. In all three methods triangular elements were used and were all arranged in the mesh pattern shown in Fig. 4. The equilibrium model II formulated by Fraeijs de Veubeke (1968) was based on assumed linear bending moment distribution within each element. The compatible model formulated by Clough and Tocher (1966) was based on cubic distribution in lateral displacement w and linear distribution in the normal slope $w_{,\nu}$ along each edge of the triangle. The hybrid stress model formulated by the present authors was based on the linear bending moment distribution within each element when the lateral displacement w is cubic and the normal slope $w_{,\nu}$ is linear along each edge. The resulting strain energy by the hybrid stress model should be bounded by that of the other two models if the element sizes are the same.

Figure 4 consists of plots of the ratios of the calculated central deflections and the exact value versus the number of elements per half span. In the present case the central deflection is a measure of the strain energy in the plate and the results are shown to follow our prediction that the hybrid model solutions are bounded by those of the other two models. Indeed, this example and other numerical experiments (Pian and Tong, 1969a; Whetstone and Yen, 1970; Wolf, 1971) have shown that the hybrid stress model is a good compromise of the equilibrium model and the compatible model and that it is more accurate than the other two models.

The discussion in the present article has been limited to equilibrium problems of linear elastic solids. There are many other problems in solid

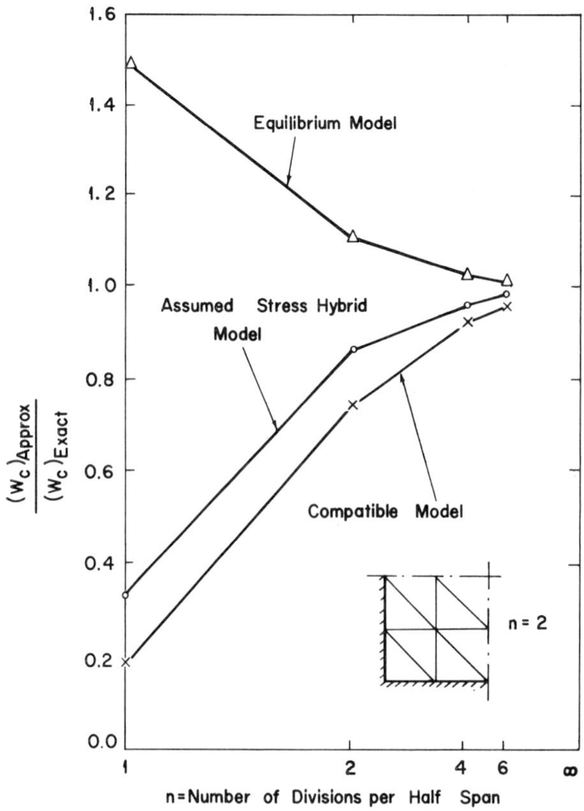

FIG. 4. Comparison of different finite element solutions for center deflection of centrally loaded square plate of clamped edges.

mechanics which can be formulated by means of variational principles and hence can be solved by finite element methods. Examples of these are the elastic stability and linear vibration analyses which are expressed in terms of eigenvalue problems and the transient response and viscoelastic analyses of structures, both of which have time as the additional variable. The finite element methods have also been extended to nonlinear problems resulting from elastic-plastic material properties or from large deflections or finite strains. A recent volume edited by Gallagher *et al.* (1971) contains survey articles on the various aspects of finite element development, and a text by Oden (1972) has a comprehensive treatment of finite element solutions of nonlinear continua. If a nonlinear problem is not path dependent, such as the deformation theory of plasticity or the prebuckling large deflection solution for a plate or a shell, the resulting nonlinear equations by the finite element formulation can be solved by various established iterative methods. Finite element solutions for nonlinear problems can also be formulated for different finite element models by an incremental approach based on the second variation of the variational functions for solid mechanics (Pian and Tong, 1971). When a nonlinear problem contains a single loading parameter for the applied load, the incremental formulation yields a system of ordinary differential equations with the loading parameter as the independent variable. Such equations can be solved by numerical integration. A large portion of the most recent work on finite element methods are analyses of shells for both linear and nonlinear solutions. For this vast subject the survey paper by Gallagher (1969) should be consulted.

III. Finite Element Formulation of Several Continuum Mechanics Problems

A. Application to Certain Heat Transfer Problems

Much work has been done on the application of the finite element method to heat conduction problems (Visser, 1966; Wilson and Nickell, 1966; Gallagher and Mallett, 1969; and many others). The governing equation for the temperature distribution in a domain V is

$$\rho\left(\frac{\partial T}{\partial t} + u_i T_{,i}\right) = (c_{ij} T_{,j})_{,i} + H, \tag{3.1}$$

and the initial condition and boundary condition are, respectively,

$$T = T_0(\mathbf{x}) \quad \text{for} \quad t = 0, \tag{3.2}$$

$$a_1 T + a_2 c_{ij} T_{,j} \nu_i = \theta \quad \text{on} \quad \partial V, \tag{3.3}$$

where ρ is the density; T, the temperature; $\mathbf{u}$, the velocity vector; c_{ij} $(= c_{ji})$, the thermal conductivity coefficient which can be a function of spatial coordinates $\mathbf{x}$, as well as the temperature itself; H, the heat source; T_0, a_1, a_2, and θ are known functions; and ν_i is the direction cosine of the unit normal over the surface ∂V. The present problem cannot be stated in the form of the stationary property of a functional; thus, in applying the finite element method, the problem will be transformed into a variational statement. We shall restrict ourselves to the problems of finite domain with a prescribed velocity field $\mathbf{u}$, and shall consider two variational statements, one of which involves only the temperature T as the field variable while the other involves $T_{,i}$ as additional variables. Thus, we can formulate two finite element models for this heat transfer problem.

1. *Model I*

We are seeking a solution such that

$$\delta\pi = \int_V \left[\left(\rho \frac{\delta T}{\delta t} + \rho u_i T_{,i} - H\right) \delta T + c_{ij} T_{,i}\, \delta T_{,j}\right] dV \\ + \int_{\partial V} \frac{1}{a_2}(a_1 T - \theta)\, \delta T\, dS = 0, \tag{3.4}$$

where the variation is for arbitrary δT. Furthermore, over the portion of the boundary where a_2 is zero, the requirement that

$$a_1 T_1 = \theta$$

and

$$\delta T = 0 \tag{3.5}$$

In formulating a finite element method the region V is, as usual, subdivided into a finite number of discrete elements V_n, each with selected node points. For each individual element the nodal values of T or also the derivatives of T are chosen as the generalized coordinates which are represented by a vector $\mathbf{q}$. Appropriate interpolations can then be constructed to approximate the temperature T over the individual elements, i.e. for the nth element,

$$T = \mathbf{f}(\mathbf{x})\mathbf{q}_n, \tag{3.6}$$

from which

$$\delta T = \mathbf{f}(\mathbf{x})\delta\mathbf{q}_n, \tag{3.7}$$

where $\mathbf{f}(\mathbf{x})$ is a row matrix of interpolation functions.

It is because the first integral in (3.4) contains a term $c_{ij} T_{,i}\, \delta T_{,j}$ that we must choose the interpolation functions $\mathbf{f}$ such that both T and δT are continuous over the entire domain V. This is to guarantee that the first integral is defined for the chosen T and δT in (3.6) and (3.7). The present formulation is, of course, applicable to one-dimensional, two-dimensional, and three-dimensional problems by using the appropriate interpolation functions described in the Appendix.

Realizing that the domain V is represented by individual elements V_n and then substituting (3.6) and (3.7) into (3.4) we obtain

$$\delta\pi = \sum_n \delta\mathbf{q}_n{}^T\left(\mathbf{m}_n \frac{d\mathbf{q}_n}{dt} + \mathbf{k}_n\mathbf{q}_n - \mathbf{Q}_n\right), \tag{3.8}$$

where $\mathbf{m}_n$, $\mathbf{k}_n$, and $\mathbf{Q}_n$ are constructed by using

$$\delta\mathbf{q}_n{}^T\mathbf{m}\frac{d\mathbf{q}_n}{dt}=\int_{V_n}\rho\frac{\partial T}{\partial t}\,\delta T\,dV,$$

$$\delta\mathbf{q}_n{}^T\mathbf{k}_n\mathbf{q}_n=\int_{V_n}(\rho u_i T_{,i}\,\delta T+c_{,j}T_{,i}\,\partial T_{,j})\,dV+\int_{\bar{S}_n}\frac{a_1}{a_2}\,T\delta T\,dS, \tag{3.9}$$

$$\delta\mathbf{q}_n{}^T\mathbf{Q}_n=\int_{\bar{S}_n}\theta\delta T\,dS+\int_{V_n}H\delta T\,dV,$$

and $\bar{S}_n$ is the portion of the boundary of V_n which is on the boundary ∂V where $a_2\neq 0$. Here again because of the common nodal points for neighboring elements, (3.8) must be transferred by a process given in Section II-B into one which contains only the global independent generalized coordinates $\mathbf{q}$.

The resulting equation is

$$\delta\pi=\delta\mathbf{q}^T\left(\mathbf{M}\frac{d\mathbf{q}}{dt}+\mathbf{K}\mathbf{q}-\mathbf{Q}\right)=0 \tag{3.10}$$

for arbitrary $\delta\mathbf{q}$. This variation, of course, also is subjected to the conditions of constraint given by (3.5) and a procedure similar to (2.20) and (2.21) should be used in the formulation. The system of algebraic equations for the determination of $\mathbf{q}$ is

$$\mathbf{M}\frac{d\mathbf{q}}{dt}+\mathbf{K}\mathbf{q}=\mathbf{Q}. \tag{3.11}$$

We shall illustrate this formulation by two examples.

Example 1. Consider a simple one-dimensional problem (Fig. 5) given by the equation

$$\rho\left(\frac{\partial T}{\partial t}+u\frac{\partial T}{\partial x}\right)=c\frac{\partial^2 T}{\partial x^2} \tag{3.12}$$

for $0\leq x\leq L$, the initial condition

$$T=T_0(x) \tag{3.13}$$

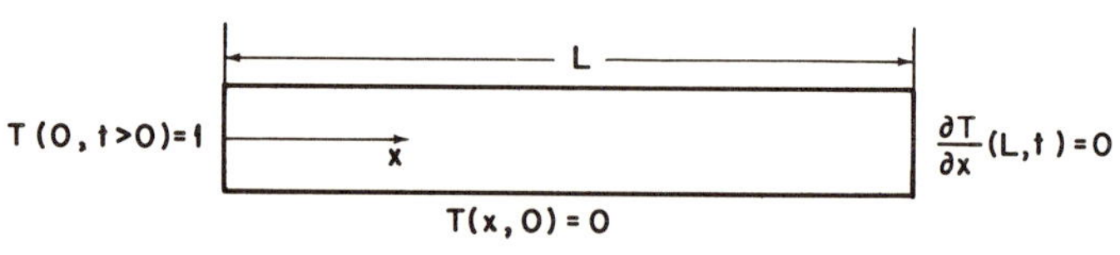

FIG. 5. One-dimensional heat transfer problem.

at $t=0$, and the boundary conditions for $t>0$

$$\begin{aligned} T&=1 \quad \text{at} \quad x=0, \\ \partial T/\partial x&=0 \quad \text{at} \quad x=L. \end{aligned} \tag{3.14}$$

Referring to (3.1) the present problem corresponds to the case where the heat source H is zero and u_1 and c_{ij} are respectively scalar constants, u and c. The boundary conditions by (3.3) are simply $a_1=1$, $a_2=0$, $\theta=1$ at $x=0$ and $a_2=1$, $a_1=\theta=0$ at $x=L$.

We divide the region into N uniform elements with length ε $(=L/N)$. The generalized coordinate q_n $(n=1, \ldots, N+1)$ is the value of T at x_n $(=(n-1)\varepsilon)$. The interpolation function is simply a linear function given in (A.4). From (3.9), the element matrices are

$$\begin{aligned} \mathbf{m}&=\rho\varepsilon\begin{bmatrix} \frac{1}{3} & \frac{1}{6} \\ \frac{1}{6} & \frac{1}{3} \end{bmatrix}, \\ \mathbf{k}&=\frac{1}{\varepsilon}\begin{bmatrix} c-\frac{1}{2}\rho u\varepsilon & -c+\frac{1}{2}\rho u\varepsilon \\ -c-\frac{1}{2}\rho u\varepsilon & c+\frac{1}{2}\rho u\varepsilon \end{bmatrix}, \end{aligned} \tag{3.15}$$

and $\bar{Q}_n$ is zero for every element. By (3.5), it is required that $q_1=1$ and $\delta q_1=0$; and (3.11) becomes

$$\frac{\rho\varepsilon}{6}(\dot{q}_{n-1}+4\dot{q}_n+\dot{q}_{n+1})=\frac{c}{\varepsilon}\left[\left(1+\frac{\rho u\varepsilon}{2c}\right)q_{n-1}-2q_n+\left(1-\frac{\rho u\varepsilon}{2c}\right)q_{n+1}\right] \quad \text{for } n=2, 3, \ldots, N$$

and

$$\frac{\rho\varepsilon}{6}(\dot{q}_N+2\dot{q}_{N+1})=\frac{c}{\varepsilon}\left[\left(1+\frac{\rho u\varepsilon}{2c}\,q_N\right)-\left(1+\frac{\rho u\varepsilon}{2c}\right)q_{N+1}\right]. \tag{3.16}$$

In practice, the values of q_n are obtained by integrating (3.16) numerically using the initial condition

$$q_n=T_0(x_n) \tag{3.17}$$

at $t=0$. In the numerical integration procedure further approximations are usually made for the left-hand sides of (3.16) and the equations are written as

$$\rho\varepsilon\dot{q}_n=\frac{c}{\varepsilon}\left[\left(1+\frac{\rho u\varepsilon}{2c}\right)q_{n-1}-2q_n+\left(1-\frac{\rho u\varepsilon}{2c}\right)q_{n+1}\right] \quad \text{for } n=2, 3, \ldots, N$$

and

$$\frac{\rho\varepsilon\dot{q}_{N+1}}{2}=\frac{c}{\varepsilon}\left[\left(1+\frac{\rho u\varepsilon}{2c}\right)q_N-\left(1+\frac{\rho u\varepsilon}{2c}\right)q_{N+1}\right]. \tag{3.18}$$

Some numerical results of (3.18) are presented in Fig. 6.

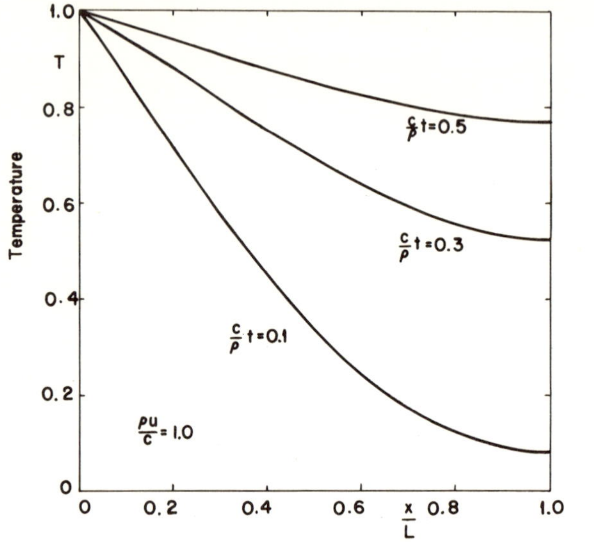

FIG. 6. Transient temperature distribution of the one-dimensional heat transfer problem.

Example 2. We consider next the steady state heat conduction in axially symmetric Stokes flow (Fig. 7) for which the heat source is zero and the thermal conductivity coefficient is a scalar constant c (Tong and Friedmann, 1972). The governing equation is

$$u_r \frac{\partial T}{\partial r} + u_z \frac{\partial T}{\partial z} = c \left[\frac{1}{r} \frac{\partial}{\partial r} \left(r \frac{\partial T}{\partial r} \right) + \frac{\partial^2 T}{\partial z^2} \right] \tag{3.19}$$

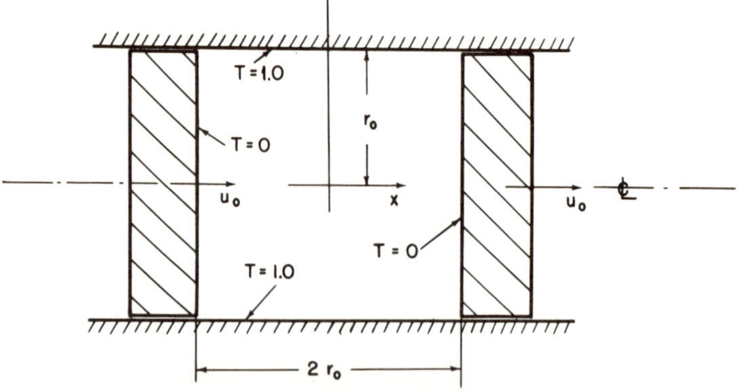

FIG. 7. Steady state heat conduction in axially symmetric Stokes flow.

and the boundary conditions are

$$\begin{aligned} T &= 0 \quad \text{at} \quad z = \pm r_0, \\ T &= 1 \quad \text{at} \quad r = r_0, \\ \partial T/\partial r &= 0 \quad \text{at} \quad r = 0. \end{aligned} \tag{3.20}$$

Referring to the boundary condition (3.3) these boundary conditions correspond to $a_2 = 0$, $a_1 = \theta = 1$ at $r = r_0$, $a_2 = 1$, $a_1 = \theta = 0$ at $r = 0$ and $a_2 = \theta = 0$, $a_1 = 1$ at $z = \pm r_0$.

For the finite element formulation we divide the region into rectangular ring elements with the four circular edges as the nodes. Since the problem is axially symmetric only the cross section r–z plane is to be considered. The appropriate interpolation function is given in (A.7). The variational equation is of the form

$$\delta\pi = \sum_n \delta\mathbf{q}_n{}^T\mathbf{k}_n\mathbf{q}_n = 0 \tag{3.21}$$

where the element matrix $\mathbf{k}_n$ is defined by

$$\delta\mathbf{q}_n{}^T\mathbf{k}_n\mathbf{q}_n = 2\pi \int_{z1}^{z2}\int_{r1}^{r2}\left[\left(u_r\frac{\partial T}{\partial r} + u_z\frac{\partial T}{\partial z}\right)\delta T + c\left(\frac{\partial T}{\partial r}\frac{\partial\delta T}{\partial r} + \frac{\partial T}{\partial z}\frac{\partial\delta T}{\partial z}\right)\right] r\,dr\,dz \tag{3.22}$$

In (3.22) u_r and u_z can be obtained for example, from the finite element solution of slow viscous flow developed by Tong and Fung (1971). The variational statement (3.21) can again be expressed in terms of independent coordinates $\mathbf{q}$ subjecting to the conditions of constraint at $z = \pm r_0$ and $r = r_0$, and then by realizing that $\delta\mathbf{q}$ is arbitrary, we can obtain a system of algebraic equations for the determination of $\mathbf{q}$. Some numerical results are given in Fig. 8.

2. *Model II*

Equations (3.1) through (3.3) can be rewritten when additional variables h_i defined by

$$h_i = T_{,i} \tag{3.23}$$

are introduced. The governing equations are then given by (3.23) and

$$\rho\left(\frac{\partial T}{\partial t} + u_i h_i\right) = (c_{ij}h_j)_{,i} + H, \tag{3.24}$$

subject to the initial condition

$$T = T_0(\mathbf{x}) \tag{3.25}$$

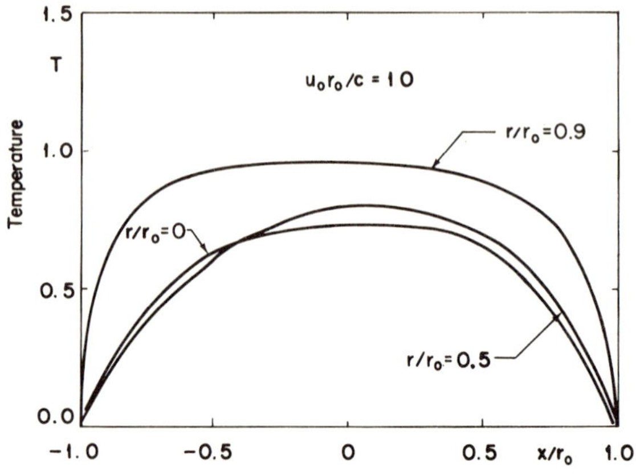

FIG. 8. Steady state temperature distribution.

at $t=0$ and the boundary condition

$$a_1 T + a_2 c_{ij} h_j \nu_i = \theta \quad \text{on} \quad \partial V. \tag{3.26}$$

A finite element solution can be constructed for this version by using the variational statement

$$\delta\pi = \int_V \left\{ \left[\rho \frac{\partial T}{\delta t} + \rho u_i h_i - H \right] \delta T + c_{ij}[h_i \, \delta T_{,j} + T_{,j} \, \delta h_i - h_j \, \delta h_i] \right\} dV$$

$$+ \int_{\partial V} \frac{1}{a_2} [a_1 T - \theta] \, \delta T \, dS = 0 \tag{3.27}$$

for arbitrary δT and δh_i. The generalized coordinates and the interpolation functions for both T and h_i can be assumed independently. The only restriction is that either T or h_i must be continuous over the entire domain in order to have the integral defined. Of course, (3.5) must also be satisfied if $a_2 = 0$ on ∂V. Some examples of the one-dimensional problem, assuming T to be constant and h_i to be linear within each element, are given by Thompson and Chen (1970).

B. Application to Steady State Temperature Distribution

The problems of steady state temperature distribution in a medium of thermal conductivity and without heat convection and heat source is a special case of the problem discussed in the previous subsection. Equation (3.1) now reduces to

$$T_{,ii} = 0 \quad \text{in} \quad V. \tag{3.28}$$

The boundary conditions are

$$T = \bar{\theta} \quad \text{on} \quad \partial\bar{V}, \tag{3.29}$$

$$a_1 T + a_2 \frac{\partial T}{\partial \nu} = \theta \quad \text{on} \quad \partial V - \partial\bar{V}, \tag{3.30}$$

where $\partial\bar{V}$ corresponds to the region where $a_2 = 0$. The variational statement (3.4) can be expressed as the minimization of

$$\pi = \frac{1}{2}\int_V (T_{,i})^2\, dV + \int_{\partial V - \partial\bar{V}} \frac{T}{a_2} (\tfrac{1}{2} a_1 T - \theta)\, dS \tag{3.31}$$

with respect to T subjected to the condition of constraint (3.29).

We shall consider an alternative way of solving (3.28) and hence present another possible finite element formulation of the problem.

It is known (Kellog, 1929) that the solution of (3.28) can be represented by

$$T(\xi) = \int_{\partial V} \frac{\Xi(\eta)}{r(\xi, \eta)}\, dS(\eta)1 \tag{3.32}$$

where ξ is a point in V or on ∂V, η is a point on ∂V, and $r(\xi, \eta)$ is the distance between ξ and η. The outward normal derivative of T on ∂V is

$$v(\xi) = \left(\frac{\partial T}{\partial \nu}\right)_\xi = 2\pi\Xi(\xi) + \int_{\partial V} \Xi(\eta) \frac{\partial}{\partial \nu}\left(\frac{1}{r(\xi, \eta)}\right) dS(\eta). \tag{3.33}$$

The function Ξ can be identified as a source distribution on ∂V; it is, of course, a function of the surface coordinates (α, β).

By using the integral representation (3.32) for T, (3.31) is reduced to

$$\pi = \frac{1}{2}\int_{\partial V} \frac{\partial T}{\partial \nu} T\, dS + \int_{\partial V - \partial\bar{V}} \left(\frac{1}{2}\frac{a_1 T^2}{a_2} - \frac{1}{a_2}\theta T\right) dS. \tag{3.34}$$

In the finite element formulation first we divide the surface ∂V into a finite number of elements and interpolate the function Ξ in terms of its nodal values, $\mathbf{q}$, i.e.

$$\Xi(\eta) = \mathbf{f}(\eta)\mathbf{q}_n \tag{3.35}$$

for the nth element. We seek next the relation between the values of T and $\partial T/\partial \nu$ at the nth node on ∂V and the values of q's. For each element n, we construct two element row vectors $\mathbf{a}_n$ and $\mathbf{b}_n$ for the mth node according to

$$\begin{aligned} \mathbf{a}_n &= \int_{\partial V_n} \frac{\mathbf{f}(\eta)}{r(\xi_m, \eta)}\, dS(\eta), \\ \mathbf{b}_n &= \int_{\partial V_n} \mathbf{f}(\eta) \frac{\partial}{\partial \nu}\left(\frac{1}{r(\xi_m, \eta)}\right) dS(\eta), \end{aligned} \tag{3.36}$$

It should be noted that both integrands in (3.36) are singular; hence special care must be taken in carrying out the integrations when $\eta \cong \xi_m$.

The values of T and $\partial T/\partial \nu$ at the mth node can thus be written as

$$
\begin{aligned}
T_m &= \sum_n \mathbf{a}_n \mathbf{q}_n = \sum_l A_{ml} q_l, \\
v_m &= (\partial T/\partial \nu)_m = \sum_n \mathbf{b}_n \mathbf{q}_n + 2\pi q_m = \sum_l B_{ml} q_l,
\end{aligned}
\tag{3.37}
$$

where q_l is the value of Ξ at the lth node, $\sum_n$ denotes a summation over all elements on the surface ∂V_n and $\sum_l$ refers to summation over all nodes on the surface. The coefficients A_{ml} and B_{ml} are obtained respectively by assembling the corresponding elements in $\mathbf{a}_n$ and $\mathbf{b}_n$. In matrix form (3.37) can be written as

$$
\begin{aligned}
\mathbf{T} &= \mathbf{Aq}, \\
\mathbf{v} &= \mathbf{Bq}.
\end{aligned}
\tag{3.38}
$$

We interpolate, then, the function T and its normal derivatives $\partial T/\partial \nu$ for each element in terms of their nodal values T_m and v_m by using the same interpolation functions $f(\eta)$, i.e. for the nth element,

$$
\begin{aligned}
T(\eta) &= \mathbf{f}(\eta)\mathbf{T}_n, \\
\partial T(\eta)/\partial \nu &= \mathbf{f}(\eta)\mathbf{v}_n.
\end{aligned}
\tag{3.39}
$$

A substitution of (3.39) into (3.34) thus yields the following expression for the function π;

$$
\pi = \sum_n (\tfrac{1}{2}\mathbf{v}_n{}^T \mathbf{c}_n \mathbf{T}_n + \tfrac{1}{2}\mathbf{T}_n{}^T \mathbf{d}_n \mathbf{T}_n - \mathbf{T}_n{}^T \bar{\mathbf{Q}}_n), \tag{3.40}
$$

where

$$
\begin{aligned}
\mathbf{c}_n &= \int_{\partial V_n} \mathbf{f}^T \mathbf{f}\, dS, \\
\mathbf{d}_n &= \int_{\partial V_n - \partial \bar{V}_n} \frac{a_1}{a_2} \mathbf{f}^T \mathbf{f}\, dS, \\
\bar{\mathbf{Q}}_n &= \int_{\partial V_n - \partial \bar{V}_n} \frac{a_1}{a_2} \theta \mathbf{f}\, dS,
\end{aligned}
\tag{3.41}
$$

where $\partial \bar{V}_n$ is the portion of ∂V_n where a_2 is zero.

Equation (3.40) is then rewritten in terms of the global values of $\mathbf{T}$ and $\mathbf{v}$ in the form

$$
\pi = \tfrac{1}{2}\mathbf{v}^T \mathbf{CT} + \tfrac{1}{2}\mathbf{T}^T \mathbf{DT} - \mathbf{T}\bar{\mathbf{Q}}, \tag{3.42}
$$

and then by substituting (3.38), π becomes

$$\pi = \tfrac{1}{2}\mathbf{q}^T(\mathbf{B}^T\mathbf{C} + \mathbf{A}^T\mathbf{D})\mathbf{A}\mathbf{q} - \mathbf{q}^T\mathbf{A}^T\bar{\mathbf{Q}}. \tag{3.43}$$

In minimizing π, the unknowns $\mathbf{q}$ must be subjected to the condition of constraint given by (3.29). By partitioning $\mathbf{T}$ in (3.38) into

$$\mathbf{T} = \begin{bmatrix} \mathbf{T} \\ \bar{\boldsymbol{\theta}} \end{bmatrix} = \begin{bmatrix} \mathbf{A}_1 \\ \mathbf{A}_2 \end{bmatrix} \mathbf{q}, \tag{3.44}$$

where $\bar{\boldsymbol{\theta}}$ is the column vector for the nodal value of θ on $\partial \bar{V}$, then (3.29) becomes

$$\mathbf{A}_2\mathbf{q} = \bar{\boldsymbol{\theta}} \tag{3.45}$$

which is the condition of constraint for $\mathbf{q}$ in (3.43).

C. Application to Two-Dimensional or Axial Symmetric Stokes Flow

In the case of slow motion of fluid the inertia force can be neglected as compared to the viscous force. The governing equation becomes simply, in the domain V,

$$\nabla \cdot \mathbf{v} = 0 \tag{3.46}$$

and

$$\mu \nabla \cdot \nabla \mathbf{v} - \nabla p = 0, \tag{3.47}$$

where $\mathbf{v}$ is the velocity vector, p the pressure, and μ the coefficient of viscosity.

We shall show that in the case of two-dimensional flow or axial symmetric flow, a different finite element approach can be used. Let us introduce the stream function ψ, defined by

$$\begin{aligned} v_1 &= \frac{1}{{x_2}^m} \frac{\partial \psi}{\partial x_2}, \\ v_2 &= -\frac{1}{{x_2}^m} \frac{\partial \psi}{\partial x_1}, \end{aligned} \tag{3.48}$$

where $m = 0$ and 1, respectively, for two-dimensional flow and axial symmetric flow. It is seen that (3.46) is satisfied identically. By eliminating p, (3.47) is reduced to

$$L({x_2}^m L\psi) = 0, \tag{3.49}$$

where

$$L = \frac{\partial}{\partial x_2}\left(\frac{1}{{x_2}^m} \frac{\partial}{\partial x_2}\right) + \frac{1}{{x_2}^m} \frac{\partial^2}{\partial {x_1}^2} \tag{3.50}$$

and x_1 is the axis of symmetry in the case of axisymmetric flow. The boundary value of ψ and $\partial\psi/\partial\nu$ is related to that of $\mathbf{v}$ by

$$\psi = \int^s x_2{}^m v_\nu \, ds$$
$$\partial\psi/\partial\nu = -x_2{}^m v_s \tag{3.51}$$

on ∂A, where v_ν and v_s are the normal and tangential components of $\mathbf{v}$, A is the cross section area of the domain considered, and ∂A is the boundary curve of A. (For a multiple connected region, see Tong and Fung, 1971.)

Equation (3.49) is similar to the plate bending problem in solid mechanics; thus one can follow the same procedure as in Section II to construct a finite element solution for ψ. Atkinson *et al.* (1969) have solved the entry flow problem by the finite element method by using triangular elements.

In the following three finite element models are described: one is a mixed model similar to that formulated by using the Reissner's variational principle and the other two are analogous to the hybrid model.

1. *Mixed Model*

We first define the vorticity, ω, by

$$\omega = \frac{\partial v_2}{\partial x_1} - \frac{\partial v_1}{\partial x_2} = L\psi. \tag{3.52}$$

The governing equations are now

$$L\psi = \omega$$

and

$$L(x_2{}^m\omega) = 0. \tag{3.53}$$

For simplicity, we shall only consider the boundary conditions on ∂A,

$$\psi = \bar{\psi}, \tag{3.54}$$

$$\partial\psi/\partial\nu = \bar{\psi}_1. \tag{3.55}$$

The corresponding variational principle for (3.53) through (3.55) is simply

$$\pi = \frac{1}{2}\int_A \left[-x_2{}^m\omega^2 + 2\nabla(x_2{}^m\omega) \cdot \frac{\nabla\psi}{x_2{}^m} \right] dx_1\, dx_2 - \int_{\partial A} \bar{\psi}_1 \omega \, ds. \tag{3.56}$$

One can easily construct a finite element solution for ψ and ω similar to that of the plane stress or plane strain problem. Since, in (3.56), the first partial derivatives of both ψ and ω appear explicitly in the finite element formulation, the assumed interpolation functions for both ψ and ω must be continuous.

2. *Hybrid Model I*

If the domain A is subdivided into discrete elements and ω is chosen so that (3.53) is identically satisfied within each individual element, (3.56) can be reduced to

$$\pi = \sum_n \left[\int_{\partial A_n} \frac{(x_2{}^m \omega),_\alpha \nu_\alpha \psi \, ds}{x_2{}^m} - \frac{1}{2} \int_{A_n} x_2{}^m \omega^2 \, dx_1 \, dx_2 \right] - \int_{\partial A} \bar{\psi}_1 \omega \, ds. \tag{3.57}$$

The procedure for constructing the finite element solution is similar to that of Section II,F. For example, one can use the nodal value of ψ or $\psi/x_2{}^m$ as generalized coordinates and, within each element, ω is assumed to be of the form

$$\begin{aligned} \omega &= \beta_1 + \beta_2 x_1 + \beta_3 x_2 \qquad &&\text{(for two-dimensional flow)}, \\ &= x_2(\beta_1 + \beta_2 x_1) &&\text{(for axial symmetric flow)}. \end{aligned} \tag{3.58}$$

3. *Hybrid Model II* (*Tong and Fung, 1971*; *Tong and Vawter, 1972*)

Let us introduce

$$\begin{aligned} M_{11} &= \frac{1}{x_2{}^m} \frac{\partial^2 \psi}{\partial x_1{}^2}, \\ M_{22} &= \frac{\partial}{\partial x_2} \left(\frac{1}{x_2{}^m} \frac{\partial \psi}{\partial x_2} \right), \\ M_{12} &= \frac{1}{x_2{}^m} \frac{\partial^2 \psi}{\partial x_1 \, \partial x_2}, \\ Q_1 &= \frac{\partial M_{11}}{\partial x_1} + \frac{\partial M_{12}}{\partial x_2}, \\ Q_2 &= \frac{\partial M_{12}}{\partial x_1} + \frac{1}{x_2{}^m} \frac{\partial (x_2{}^m M_{22})}{\partial x_2}. \end{aligned} \tag{3.59}$$

Then (3.49) is equivalent to

$$\frac{\partial^2 M_{11}}{\partial x_1{}^2} + 2 \frac{\partial^2 M_{12}}{\partial x_1 \, \partial x_2} + \frac{\partial}{\partial x_2} \left[\frac{1}{x_2{}^m} \frac{\partial (x_2{}^m M_{22})}{\partial x_2} \right] = 0 \tag{3.60}$$

and

$$\begin{aligned} \frac{1}{x_2{}^m} \frac{\partial (x_2{}^m M_{11})}{\partial x_2} &= \frac{\partial M_{12}}{\partial x_1} \\ \frac{\partial M_{22}}{\partial x_1} &= \frac{\partial M_{12}}{\partial x_2}. \end{aligned} \tag{3.61}$$

It can be shown that the present Stokes flow problem is equivalent to a variational principle with the following expression for the functional:

$$\pi = \sum_n \left[\int_{\partial A_n} (M_{\alpha\beta} \nu_\alpha \psi_{,\beta} - Q_\alpha \nu_\alpha \psi)\, ds - \frac{1}{2} \int_{A_n} (M_{11}^2 + 2M_{12}^2 + M_{22}^2) x_2{}^m \, dx_1 \, dx_2 \right], \qquad (3.62)$$

where $\sum_n$ denotes the sum over all elements. The procedure of constructing the finite element solution is the same as that of Section II,F. Incidentally, in the case of two-dimensional flow, the governing equations are similar to that of the plate bending. In the paper by Tong and Fung (1971) the M's are assumed in the form

$$M_{11} = \beta_1 + \beta_2 x_1 + \beta_3 x_2 + \beta_4 x_1{}^2 + \beta_5 x_1 x_2 + \beta_6 x_2{}^2,$$

$$M_{12} = \beta_7 + \beta_8 x_1 + \beta_9 x_2 + \beta_{10} x_1{}^2 + \beta_{11} x_1 x_2 + \beta_{12} x_2{}^2,$$

$$M_{22} = \begin{cases} \beta_{13} + \beta_{14} x_1 + \beta_{15} x_2 + \beta_{16} x_1{}^2 + \beta_{17} x_1 x_2 - (\beta_4 + \beta_{11}) x_2{}^2 \\ \qquad \text{(for two-dimensional flow)} \\ x_2(\beta_{13} + \beta_{14} x_1 + \beta_{15} x_1{}^2) - \frac{2}{3}(\beta_4 + \beta_{11}) x_2 \\ \qquad \text{(for axial symmetric flow)} \end{cases} \qquad (3.63)$$

and a new variable ϕ is used for $\psi/x_2{}^m$. The nodal values of ϕ, $\partial\phi/\partial x_1$, and $\partial\phi/\partial x_2$ are then used as generalized coordinates.

D. Remarks

Many other problems in continuum mechanics can be solved by the finite element method similar to the one described in this section. For example, the lubrication problems are governed by the Reynolds equation which corresponds to the case when $\partial T/\partial t = 0$ in (3.1) (Reddi, 1969; Reddi and Chu, 1970). The potential flow problem is governed by a Laplace equation. Tong (1966), Martin (1968), and Luk (1969) have used the approaches given in Section III,A. Guyan *et al.* (1969) and Tong (1971) have discussed the use of the integral equation approach for the liquid sloshing problems. In incompressible viscous flow, one can easily express the continuity equation and the momentum equations in the form of a variational statement similar to (3.4), and then construct a finite element solution accordingly (Tong, 1971; Skalak *et al.*, 1971). Problems in aerodynamics (Kariappa, 1970) and in aeroelasticity (Olson, 1970; Kariappa and Somashaka, 1969; Tong *et al.*, 1970) have also been solved by the finite element methods.

Appendix. Interpolation Functions

One significant step in the finite element formulations is to approximate the field variable by a set of assumed piecewise continuous and differentiable shape functions defined in terms of their nodal values or, in many cases, also the derivatives at the nodal points. Let $\mathbf{x} = (x_1, x_2, \ldots, x_n)$ be a point in an n-dimensional Euclidean space, then a function $u(\mathbf{x})$ within a domain D can be approximated by

$$u(\mathbf{x}) = \sum_{i=1}^{M} \phi_i(\mathbf{x}) q_i = \boldsymbol{\phi}(\mathbf{x})\mathbf{q}, \tag{A.1}$$

where q_i is either the value of u or a derivative of u at some discrete point in the domain and, in the terminology of the numerical analysis, $\boldsymbol{\phi}(\mathbf{x})$ are called the interpolation functions.

One of the general interpolation schemes for one-dimensional space is the Hermite interpolation. An approximation $\bar{u}(x)$ of a function $u(x)$ is expressed in terms of the values of $u(x)$ and derivatives (up to the Nth) of $u(x)$ at m discrete stations by

$$\bar{u}(x) = \sum_{k=1}^{m} [H_{0k}^{(N)}(x)u(x_k) + H_{lk}^{(N)}(x)u'(x_k) + H_{Nk}^{(N)}(x)u^{(N)}(x_k)]. \tag{A.2}$$

where $H_{jk}^{(N)}(x)$ is called the Hermite polynomial. The total number of terms in the Hermite interpolation formula is $M = m(N+1)$. The Hermite polynomials have the following properties:

$$\left.\frac{d^l H_{kj}^{(N)}}{dx^l}\right|_{x_i} = \delta_{ij}\,\delta_{lk} \qquad \begin{array}{l} i, j = 1, 2, \ldots, m \\ l, k = 0, 1, 2, \ldots, N \end{array} \tag{A.3}$$

where δ_{ij} is the Kronecker delta. One can easily verify that the Hermite polynomial $H_{jk}^{(N)}(x)$ is of degree $m(N+1) - 1$ or less. For example, when $N = 0$ and $m = 2$ with the two stations located at $x = 0$ and l, the Hermite polynomials, which correspond to the linear interpolation, are simply

$$\begin{aligned} H_{01}^{(0)}(x) &= 1 - (x/l), \\ H_{02}^{(0)}(x) &= x/l. \end{aligned} \tag{A.4}$$

When $N = 0$ and $m = 3$ with the three stations located at $\xi = -1$, 0, and $+1$, the Hermite polynomials are quadratic.

$$\begin{aligned} H_{01}^{(0)}(\xi) &= -\tfrac{1}{2}\xi(1-\xi), \\ H_{02}^{(0)}(\xi) &= (1-\xi)(1+\xi), \\ H_{03}^{(0)}(\xi) &= +\tfrac{1}{2}\xi(1+\xi). \end{aligned} \tag{A.5}$$

Similar cubic Hermite polynomials can be constructed to interpolate a function over four consecutive stations. When $N=1$ and $m=2$ with the two stations located at $x=0$ and l, the Hermite polynomials are also cubic functions,

$$\begin{aligned}
H_{01}^{(1)}(x) &= 2(x/l)^3 - 3(x/l)^2 + 1, \\
H_{02}^{(1)}(x) &= -2(x/l)^3 + 3(x/l)^2, \\
H_{11}^{(1)}(x) &= [(x/l)^3 - 2(x/l)^2 + (x/l)]l, \\
H_{12}^{(1)}(x) &= [(x/l)^3 - (x/l)^2]l.
\end{aligned} \tag{A.6}$$

For the two-dimensional space, a bilinear interpolation can be easily constructed for a rectangular element of dimension $a \times b$ (Fig. 9a).

$$\bar{u}(x, y) = \sum_{i=1}^{2} \sum_{j=1}^{2} H_{0i}^{(0)}(x) H_{0j}^{(0)}(y) u(x_i, y_j), \tag{A.7}$$

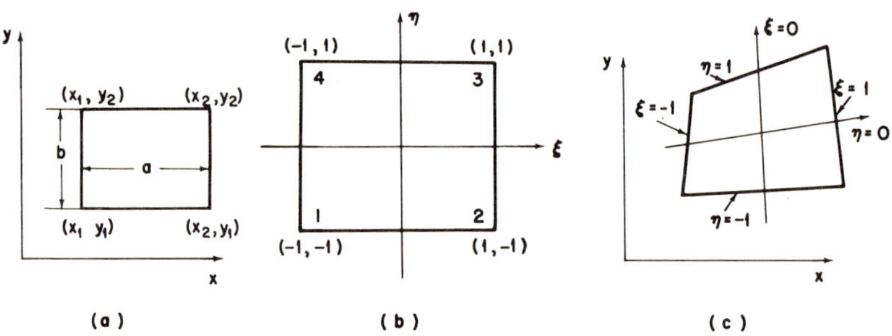

FIG. 9. Bilinear interpolation for quadrilateral element.

where for $H_{0i}(x)$, $l=a$, and for $H_{0j}(y)$, $l=b$. If the boundary of the rectangular element is given by $\xi = \pm 1$ and $\eta = \pm 1$ and the corners are labeled 1 to 4 as shown in Fig. 9b, then

$$\bar{u}(\xi, \eta) = \sum_{i=1}^{4} p_i(\xi, \eta) u_i, \tag{A.8}$$

where

$$p_i(\xi, \eta) = \tfrac{1}{4}(1 + \xi\xi_i)(1 + \eta\eta_i). \tag{A.9}$$

In a similar manner a biquadratic function over a rectangular element can be expressed in terms of the values of the function at nodes which are the four corner points and the midpoints of the four sides, and a bicubic function can be expressed in terms of values of the function at twelve boundary nodes.

It is seen that (A.7) can also be used for a general quadrilateral element with straight edges shown in Fig. 9c if an isoparametric transformation

(Irons, 1966) is used between the x–y coordinates and the ξ–η coordinates. Such a transformation is, in fact, simply

$$x = \sum_{i=1}^{4} p_i(\xi, \eta) x_i, \qquad y = \sum_{i=1}^{4} p_i(\xi, \eta) y_i. \tag{A.10}$$

The isoparametric transformation can be similarly extended to quadrilateral elements with curved edges of parabolic and cubic distributions by using, respectively, the biquadratic or bicubic interpolation (Ergatoudis *et al.*, 1968). It should also be noted that when the bilinear interpolations, the biquadratic, or bicubic interpolations are used to represent the function $u(x, y)$ in the individual elements of a two-dimensional domain, the following two conditions hold: (a) the continuity of $u(x, y)$ with a neighboring element over which the same interpolation scheme is used, (b) with appropriate combinations of the nodal values u_i, the constant state and linear distributions for $u_i(x, y)$ can be achieved. The latter condition implies that if these interpolation formulas are expressed in terms of a power series of expansions the constant and linear terms always exist. It was indicated in Section II that this latter condition is essential to the convergence of the finite element solution of a problem which contains $u_{,x}{}^2$, $u_{,x}$, $u_{,y}$, and $u_{,y}{}^2$ in the variation functional. Such a requirement is referred to as the completeness requirement.

Another bicubic interpolation is to make use of the cubic Hermite functions given by (A.6); hence to include not only the values of the function at the nodal points but also their derivatives as the nodal parameters. Such an interpolation on a rectangular element bounded by x_1, x_2, y_1 and y_2 is given by

$$\begin{aligned} \bar{u}(x, y) = \sum_{i=1}^{2} \sum_{j=1}^{2} [&H_{0i}^{(1)}(x) H_{0j}^{(1)}(y) u(x_i, y_j) \\ &+ H_{1i}^{(1)}(x) H_{0j}^{(1)}(y) u_{,x}(x_i, y_j) \\ &+ H_{0i}^{(1)}(x) H_{1j}^{(1)}(y) u_{,y}(x_i, y_j) \\ &+ H_{1i}^{(1)}(x) H_{1j}^{(1)}(y) u_{,xy}(x_i, y_j)]. \end{aligned} \tag{A.11}$$

It is seen that for this interpolation not only the first derivatives $u_{,x}$ and $u_{,y}$ at the corners are used; the cross second derivatives $u_{,xy}$ are also included, An interpolation formula obtained by eliminating the $u_{,xy}(x_i, y_j)$ terms from (A.11) is, of course, also a bicubic interpolation. It turns out that the interpolation formula (A.11) contains all of the constant, linear, and quadratic terms of the polynomial expansion; hence it satisfies the completeness requirement for the finite element solution for problems which contain the products of the second derivatives of u in the variational functional. On the other hand if the $u_{,xy}(x_i, y_i)$ terms are absent, the xy term will not appear in the polynomial expansion and the completeness requirement will not be fulfilled. It can also be easily verified that the interpolation formula (A.11) will maintain the compatibility of the function u and the normal derivatives $u_{,\nu}$ along the interelement boundaries. For example, the normal derivative $u_{,x}(y)$ at the boundary $x =$ constant is a cubic function of y and is interpolated by $u_{,x}$ and $u_{,xy}$ at the two ends of that boundary. It should be noted, however, that if a similar isoparametric transformation is used to extend this interpolation function to a general quadrilateral element the compatibility of the normal derivative $u_{,\nu}$ along the interelement boundaries will, in general, not be maintained.

The most versatile element in the finite element formulation of the two-dimensional problem is the triangular element. By an assemblage of triangular elements a domain of any boundary shape can be approximated.

A function $u(x, y)$ may be expanded into a polynomial

$$\begin{aligned} u(x, y) = \alpha_1 + \alpha_2 x + \alpha_3 y + \alpha_4 x^2 + \alpha_5 xy + \alpha_6 y^2 \\ + \alpha_7 x^3 + \alpha_8 x^2 y + \alpha_9 xy^2 + \alpha_{10} y^3 + \cdots. \end{aligned} \tag{A.12}$$

Thus for a simple linear interpolation three parameters are needed, and over a triangular element such an interpolation may be in terms of values of the variable at the three corner nodes. Similarly a quadratic interpolation will consist of six nodes, three at the corners and three at the midpoints of the sides. A cubic interpolation will consist of ten nodes, nine of which are along the boundary and one of which is located inside the element.

If the three corners of a triangular element which are labeled as 1, 2, and 3 are located at (x_1, y_1), (x_2, y_2), and (x_3, y_3), respectively, a linear interpolation formula can be easily constructed as

$$u(x, y) = \sum_i f_i u(x_i, y_i), \tag{A.13}$$

where

$$\begin{aligned} f_1 &= \frac{\det \begin{vmatrix} x & y & 1 \\ x_2 & y_2 & 1 \\ x_3 & y_3 & 1 \end{vmatrix}}{2A} \\ f_2 &= \frac{\det \begin{vmatrix} x_1 & y_1 & 1 \\ x & y & 1 \\ x_3 & y_3 & 1 \end{vmatrix}}{2A} \\ f_3 &= \frac{\det \begin{vmatrix} x_1 & y_1 & 1 \\ x_2 & y_2 & 1 \\ x_3 & y_3 & 1 \end{vmatrix}}{2A}, \end{aligned} \tag{A.14}$$

and A = area of the triangle

$$=\frac{1}{2}\det\begin{vmatrix} x_1 & y_1 & 1 \\ x_2 & y_2 & 1 \\ x_3 & y_3 & 1 \end{vmatrix}.$$

In expressing the interpolation function over an arbitrary triangle, it is most convenient to use the so-called triangular or natural coordinate system. To locate any point p within a triangle we consider the three subtriangles defined by p (Fig. 10). The triangular coordinates ζ_i ($i = 1, 2, 3$) are defined by

$$\zeta_i = A_i/A. \tag{A.15}$$

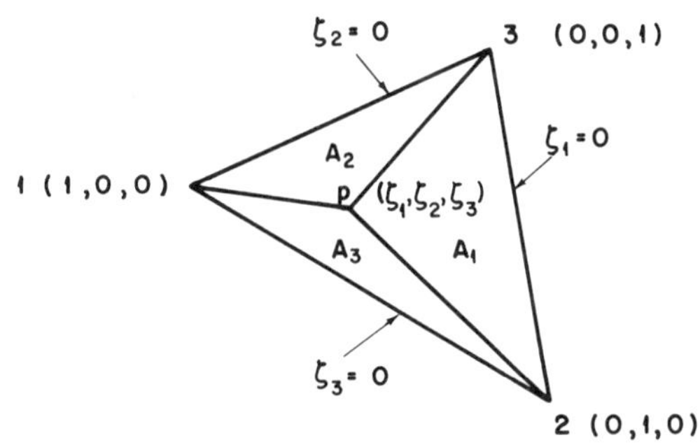

FIG. 10. Triangular coordinates for triangular element.

The triangular coordinates are thus also called the area coordinates and the following relation exists among the three coordinates$_\alpha$

$$\zeta_1 + \zeta_2 + \zeta_3 = 1. \tag{A.16}$$

In terms of the triangular coordinates the linear interpolation over a triangle is simply

$$u(\zeta_i) = \sum_{i=1}^{3} u(x_i, y_i)\zeta_i, \tag{A.17}$$

and if the boundary nodes are labeled according to Fig. 11, the quadratic interpolation is given by

$$u(\zeta_i) = [u_1\ u_2\ u_3\ u_4\ u_5\ u_6]\begin{bmatrix} \zeta_1(2\zeta_1 - 1) \\ \zeta_2(2\zeta_2 - 1) \\ \zeta_3(2\zeta_3 - 1) \\ 4\zeta_1\zeta_2 \\ 4\zeta_2\zeta_3 \\ 4\zeta_3\zeta_1 \end{bmatrix}.$$

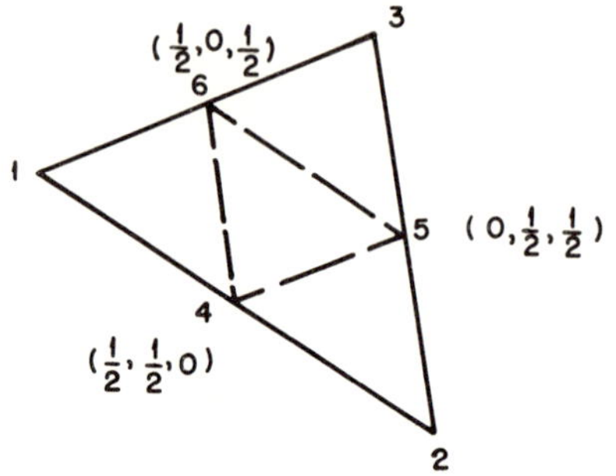

FIG. 11. Triangular elements using quadratic interpolation.

Isoparametric transformations can also be used to construct the interpolation formulas of triangular elements with curved sides. More detailed discussions of the interpolation functions over triangular elements can be found in the article by Felippa and Clough (1970).

The linear interpolation over an area which is decomposed into triangular elements can also be depicted by Fig. 12. In (A.1) if $q_i = u_i$, then the interpolation function $\phi_i(x, y)$ is a pyramid type function, i.e. it is equal to unity at the node i and slopes linearly toward the opposite edges of the triangles around this node. The function $\phi_i(x, y)$ remains zero outside the area immediate to this node. The linear interpolation obviously cannot create a smooth surface.

To obtain a smooth interpolation over a surface it is necessary to use higher order functions. Unfortunately the use of a cubic interpolation

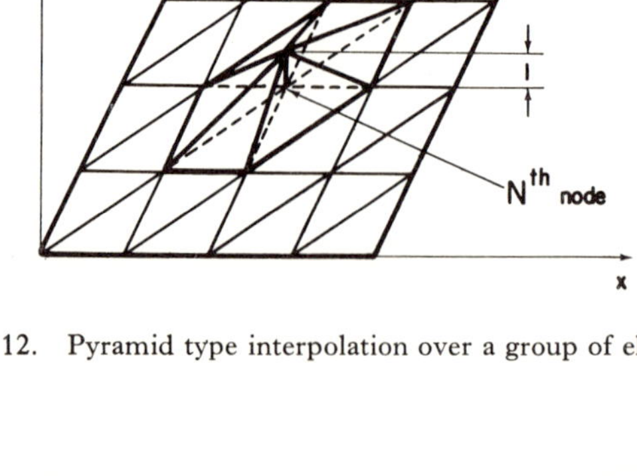

FIG. 12. Pyramid type interpolation over a group of elements.

function taking the values of the function and the two first derivatives as the generalized coordinates will, in general, only maintain the continuity of the function along the interelement boundary but not that of the normal derivatives of the function. To construct a smooth surface interpolation over a triangle or a general quadrilateral we often need to divide the element into subregions, each of which has different analytical expressions (Birkhoff and Garabedian, 1960; Clough and Tocher, 1966; Deák and Pian, 1967; Fraeijs de Veubeke, 1968). Such interpolation functions, thus, are not continuously differentiable within the element. A continuous interpolation function which maintains interelement continuity of normal derivatives can, however, be constructed when nodal values of derivatives higher than the first are used as the parameters. An example of this for a triangular element is a quintic interpolation which has 21 parameters which are the quantities u, $u_{,x}$, $u_{,y}$, $u_{,xx}$, $u_{,xy}$, $u_{,yy}$ at each of the three corners and the normal slope $u_{,\nu}$ at the midpoint of each side. This interpolation function has been used by many authors (Argyris *et al.*, 1969; Bell, 1969).

In a three-dimensional problem, the most versatile element shape is a tetrahedron. For such an element again either linear or higher order interpolations can be made, and natural coordinates based on the volumes of subtetrahedrons are the most convenient system to be used. Similar to the rectangular element for the two-dimensional problem a rectangular block element can be interpolated directly by trilinear, triquadratic, and higher order polynomials. Also, through the use of the isoparametric transformations the interpolation to a general hexahedral element with either flat faces or curved faces and a tetrahedron with curved faces can be easily constructed.

ACKNOWLEDGMENTS

The authors wish to acknowledge their appreciation to the Office of Scientific Research for their support over many years of the M.I.T. research programs (AFOSR Grant 347-65 and Contract F44620-C-0019) on methods of structural analysis which provided the essential background for the present article.

REFERENCES

Argyris, J. H. (1958). On the analysis of complex elastic structures. *Appl. Mech. Rev.* **11**, 331–338.

Argyris, J. H. (1960). "Energy Theorems and Structural Analysis." Butterworth, London (series of articles published in *Aircraft Engineering* during 1954–1955).

Argyris, J. H., Fried, I., and Scharpf, D. W. (1969). The TUBA family of elements for matrix displacement method. *J. Roy. Aeronaut. Soc.* **72**, 701–709.

Atkinson, B., Brocklebank, M. P., Card, C. C. M., and Smith, J. M. (1969). Low Reynolds number developing flows. *AIChE J.* **15**, 548–553.

Atluri, S. (1969). Static analysis of shells of revolution using doubly-curved quadrilateral elements derived from alternate variational models. Sc.D. Thesis, Dept. of Aero. and Astro., Massachusetts Institute of Technology, Cambridge.

Babuska, I. (1971). Error-bounds for finite element method. *Numer. Math.* **16**, 322–333.

Bell, K. (1969). A refined triangular plate bending finite element. *Int. J. Numer. Method Eng.* **1**, 101–122.

Berke, L., Bader, R. M., Mykytow, W. J., Przemieniecki, J. S., and Shirk, M. H., eds. (1969). "Proceedings of the Second Conference on Matrix Methods in Structural Mechanics," AFFDL-TR-68-150. Wright Patterson Air Force Base, Ohio.

Besseling, J. F. (1963). The complete analogy between the matrix equations and continuous field equations of structural analysis. *Colloq. Int. Tech. Calcul Anal. Numer. Aeronaut.*, pp. 223–242.

Birkhoff, G., and Garabedian, H. I. (1960). Smooth surface interpolation. *J. Math. Phys.* **39**, 258–268.

Birkhoff, G., Schultz, M. H., and Varga, R. S. (1968). Piecewise Hermite interpolation in one and two variables with applications to partial differential equations. *Numer. Math.* **11**, 232–256.

Clough, R. W. (1960). The finite element method in plane stress analysis. *Proc. Amer. Soc. Civil Eng.* **87**, 345–378.

Clough, R. W., and Tocher, J. L. (1966). Finite element stiffness matrices for analysis of plate bending. *Proc. Conf. Matrix Methods Struct. Mech., 1st,* 1965 AFFDL-TR-66-80, pp. 515–546.

Courant, R. (1943). Variational methods for the solution of problems of equilibrium and vibrations. *Bull. Amer. Math. Soc.* **49**, 1–23.

Deák, A. L., and Pian, T. H. H. (1967). Application of the smooth-surface interpolation to the finite-element analysis. *AIAA J.* **5**, No. 1, 187–189.

Elias, Z. M. (1967). On the duality between the problems of stretching and of bending of plates. *NASA Contract. Rep.* **NASA CR-71**.

Elias, Z. M. (1968). Duality in finite element methods. *J. Eng. Mech. Div., Amer. Soc. Civil Eng.* **94**, No. EM4, 931–946.

Ergatoudis, I., Irons, B. M., and Zienkiewicz, O. C. (1968). Curved, isoparametric 'quadrilateral' elements for finite element analysis. *Int. J. Solids Struct.* **4**, No. 1, 31–42.

Felippa, C. A., and Clough, R. W. (1970). The finite element method in solid mechanics. *In* "Numerical Solution of Field Problems in Continuum Physics" (G. Birkhoff and R. S. Varga, eds.), pp. 210–252. Amer. Math. Soc., Providence, R.I.

Fix, G., and Strang, G. (1969). Fourier analyses of the finite element method in Ritz-Galerkin theory. *Stud. Appl. Math.* **48**, 265–273.

Fox, L., and Southwell, R. V. (1945). Relaxation methods applied to engineering problems. VII. Biharmonic analysis as applied to the flexure and extension of flat elastic plates. *Phil. Trans. Roy. Soc. London, Ser. A* **239**, 419–460.

Fraeijs de Veubeke, B. (1964). Upper and lower bounds in matrix structural analysis. *AGARDograph* **72**, 165–201.

Fraeijs de Veubeke, B. (1965). Displacement and equilibrium models in the finite element method. *In* "Stress Analysis" (O. C. Zienkiewicz and G. S. Hollister, eds.), pp. 145–197. Wiley, New York.

Fraeijs de Veubeke, B. (1966). Bending and stretching of plates—special models for upper and lower bounds. *Proc. Conf. Matrix Methods Struct. Mech., 1st,* 1965 AFFDL-TR-66-80, pp. 863–886.

FRAEIJS DE VEUBEKE, B. (1968). A conforming finite element for plate bending. *Int. J. Solids Struct.* **4**, No. 1, 95–108.

FRAEIJS DE VEUBEKE, B., ed. (1971). "High Speed Computing of Elastic Structures." Proc. IUTAM Symp. 1970. Congr. Colloq. Univ. Liege.

FRAEIJS DE VEUBEKE, B., and SANDER, G. (1968). An equilibrium model for plate bending. *Int. J. Solids Struct.* **4**, No. 4, 447–468.

FRAEIJS DE VEUBEKE, B., and ZIENKIEWICZ, O. C. (1967). Strain energy bounds in finite element analysis by slab analogy. *J. Strain Anal.* **2**, No. 4, 265–271.

FUNG, Y. C. (1953). Bending of thin elastic plates of variable thickness. *J. Aeronaut. Sci.* **20**, 455–468.

GALLAGHER, R. H. (1964). "A Correlation Study of Methods of Matrix Structural Analysis." Pergamon, Oxford.

GALLAGHER, R. H. (1969). Analysis of plate and shell structures. *In* "Application of Finite Element Methods in Civil Engineering" (W. H. Rowan, Jr. and R. M. Hackett, eds.), pp. 155–205. School of Eng., Vanderbilt Univ., Nashville, Tennessee.

GALLAGHER, R. H., and MALLETT, R. (1969). Efficient solution process for finite element analysis of transient heat conduction. *Trans. ASME* Pap. 69-WA/HT-32.

GALLAGHER, R. H., YAMADA, Y., and ODEN, J. T., eds. (1971). "Recent Advances in Matrix Methods of Structural Analysis and Design." Univ. of Alabama Press, University, Alabama.

GOLDENVEIZER, A. L. (1940). The equations of the theory of thin shells. *Prikl. Mat. Mekh.* **4**, 32–42.

GREENE, B. E., JONES, R. E., MCKAY, R. W., and STROME, D. R. (1969). General variational principles in the finite element method. *AIAA J.* **7**, 1254–1260.

GUNTHER, W. (1961). Analoge Systeme von Schalen-Gleichungen. *Ing.-Arch.* **30**, 160–186.

GUYAN, R. J., UJIHARA, B. H., and WELCH, P. W. (1969). Hydroelastic analysis of axisymmetric system by a finite element method. *Proc. Conf. Matrix Methods Struct. Mech., 2nd,* 1968 AFFDL-TR-68-150, pp. 1165–1194.

HARVEY, J. W., and KELSEY, S. (1971). Triangular plate bending element with enforced compatibility. *AIAA J.* **9**, No. 6, 1023–1026.

HERRMANN, L. R. (1966). A bending analysis for plates. *Proc. Conf. Matrix Methods Struct. Mech., 1st,* 1965 AFFDL-TR-66-80, pp. 577–604.

HERRMANN, L. R. (1967). Finite element bending analysis for plates. *J. Eng. Mech. Div., Amer. Soc. Civil. Eng.* **98**, No. EM5, 13–26.

HOLAND, I., and BELL, K., eds. (1969). "Finite Element Methods in Stress Analysis." TAPIR—The Technical University of Norway, Trondheim.

IRONS, B. M. (1966). Engineering application of numerical integration in stiffness methods. *AIAA J.* **4**, 2035–2037.

JONES, R. E. (1964). A generalization of the direct-stiffness method of structural analysis. *AIAA J.* **2**, No. 5, 821–826.

KARIAPPA (1970). Kinematically consist unsteady aerodynamic coefficient in supersonic flow. *Int. J. Numer. Method Eng.* **2**, 495–507.

KARIAPPA and SOMASHAKA, B. P. (1969). Application of matrix displacement method in the study of panel flutter. *AIAA J.* **7**, 50–53.

KELLOG, O. D. (1929). "Foundation of Potential Theory." Springer-Verlag, Berlin and New York.

KEY, S. W. (1966). A convergence investigation of the direct stiffness method. Ph.D. Thesis, Dept. of Aero. and Astro., University of Washington, Seattle.

LIVESLEY, R. K. (1964). "Matrix Methods of Structural Analysis." Pergamon, Oxford.

LUK, C. H. (1969). Finite element analysis for liquid sloshing problems. S.M. Thesis, AFOSR 69-1504TR, Dept. of Aero. and Astro., Massachusetts Institute of Technology, Cambridge.

Martin, H. C. (1969). Finite element analysis of fluid flow. *Proc. Conf. Matrix Methods Struct. Mech., 2nd,* 1968 AFFDL-TR-68-150, pp. 517–538.

Melosh, R. J. (1963). Basis for derivation of matrices for the direct stiffness method. *AIAA J.* **1**, No. 7, 1631–1637.

Morley, L. S. D. (1966). Some variational principles in plate bending problems. *Quart. J. Mech. Appl. Math.* **19**, 371–386.

Morley, L. S. D. (1967). A triangular element with linearly varying bending moments for plate bending problems. *J. Roy. Aeronaut. Soc.* **71**, 715–719.

Morley, L. S. D. (1968). The triangular equilibrium element in the solution of plate bending problems. *Aeronaut. Quart.* **19**, 149–169.

Oden, J. T. (1969). A general theory of finite elements. I. Topological considerations. II. Applications. *Int. J. Num. Methods Eng.* **1**, 205–221 and 247–259.

Oden, J. T. (1972). "Finite Elements of Nonlinear Continua." McGraw-Hill, New York.

Olson, M. D. (1970). Some flutter solution, using finite elements. *AIAA J.* **8**, 747–752.

Pian, T. H. H. (1964). Derivation of element stiffness matrices by assumed stress distributions. *AIAA J.* **2**, No. 7, 1333–1336.

Pian, T. H. H. (1966). Element stiffness matrices for boundary compatibility and for prescribed boundary stresses. *Proc. Conf. Matrix Methods Struct. Mech., 1st* 1965, AFFDL-TR-66-80, pp. 457–477.

Pian, T. H. H. (1970). Finite element stiffness methods by different variational principles in elasticity. *In* "Numerical Solution of Field Problems in Continuum Physics" (G. Birkhoff and R. S. Varga, eds.), pp. 253–271. Amer. Math. Soc., Providence, R.I.

Pian, T. H. H. (1971a). Formulations of finite element methods for solid continua. *In* "Recent Advances in Matrix Methods in Structural Analysis and Design" (R. H. Gallagher, Y. Yamada, and J. T. Oden, eds.), pp. 49–83. Univ. of Alabama Press, Tuscaloosa.

Pian, T. H. H. (1971b). Variational formulations of numerical methods in solid continua. *In* "Computer-Aided Engineering" (G. M. L. Gladwell, ed.), pp. 421–448. University of Waterloo, Waterloo, Canada.

Pian, T. H. H., and Tong, P. (1969a). Rationalization in deriving element stiffness matrix by assumed stress approach. *Proc. Conf. Matrix Methods Struct. Mech., 2nd,* 1968 AFFDL-TR-68-150, pp. 441–469.

Pian, T. H. H., and Tong, P. (1969b). Basis of finite element methods for solid continua. *Int. J. Numer. Method Eng.* **1**, 3–28.

Pian, T. H. H., and Tong, P. (1971). Variational formulation of finite-displacement analysis. *In* "High Speed Computing of Elastic Structures" (B. Fraeijs de Veubeke, ed.), pp. 43–63. Congr. Colloq. Univ. Liege.

Pian, T. H. H., Tong, P. and Luk, C. H. (1972). Elastic crack analysis by a finite element hybrid method. *Proc. Conf. Matrix Methods Struct. Mech., 3rd,* 1971.

Prager, W. (1967). Variational principles of linear elastostatics for discontinuous displacements, strains and stresses. *In* "Recent Progress in Applied Mechanics" (B. Broberg, J. Hult, and F. Niordson, eds.), The Folke-Odquist Volume, pp. 463–474. Almqvist and Wiksell, Stockholm.

Prager, W. (1968). Variational principles for elastic plates with relaxed continuity requirements. *Int. J. Solids Struct.* **4**, No. 9, 837–844.

Przemieniecki, J. S., Bader, R. M., Bozich, W. F., Johnson, J. R., and Mykytow, W. J., eds. (1966). "Proceedings of the First Conference on Matrix Methods in Structural Mechanics," AFFDL-TR-66-80. Wright-Patterson Air Force Base, Ohio.

Reddi, M. M. (1969). Finite element solution of incompressible lubrication problems. *J. Lubric. Technol.* **91**, 524–533.

Reddi, M. M., and Chu, T.Y. (1970). Finite element solution of the steady-state compressible lubrication problems. *J. Lubric. Technol.* **92,** 495–503.

Reissner, E. (1950). On a variational theorem in elasticity. *J. Math. Phys.* **29**, No. 2, 90–95.

Reissner, E., and Wan, F. Y. M. (1969). On the equations of linear shallow shell theory. *Stud. Appl. Math.* **48**, No. 2, 133–145.

Rowan, W. H., Jr., and Hackett, R. M., eds. (1969). "Proceedings of the Symposium on Application of Finite Element Methods in Civil Engineering." School of Eng., Vanderbilt Univ., Nashville, Tennessee.

Skalak, R., Zarda, P. R., Chen, P. H., and Chen, T. C. (1971). A variational principle for slow viscous flow with expanded particles. *In* "Computer-Aided Engineering" (G. M. L. Gladwell, ed.), pp. 471–490. Univ. of Waterloo, Waterloo, Canada.

Sörensen, M., ed. (1969). "Finite Element Techniques." *Proc. held at Institut fur Statik und Dynamik der Luft und Raumfahrtkonstruktionen,* University of Stuttgart, Germany.

Southwell, R. V. (1950). On the analogues relating flexure and extension of flat plates. *Quart. J. Mech. Appl. Math.* **3**, 257–270.

Szabo, B. A., and Lee, G. C. (1969). Derivation of stiffness matrices for problems in elasticity by Galerkin's method. *Int. J. Numer. Methods Eng.* **1**, 301–310.

Thompson, J. J., and Chen, P. Y. P. (1970). Discontinuous finite element in thermal analysis. *Nuclear Eng. Des.* **14**, 211–222.

Tong, P. (1966). Liquid sloshing in an elastic container. Ph. D. Thesis, AFOSR 66-0943, California Institute of Technology, Pasadena, Cal.

Tong, P. (1970). New displacement hybrid finite element model for solid continua. *Int. J. Numer. Methods Eng.* **2**, 78–83.

Tong, P. (1971). The finite element method for fluid flow. *In* "Recent Advances in Matrix Methods in Structural Analysis and Design" (R. H. Gallagher, Y. Yamada, and J. T. Oden, eds.), pp. 787–808. Univ. of Alabama, Tuscaloosa.

Tong, P., and Friedmann, P. (1972). Diffusion problems in slow particulate flow in tube (to be published).

Tong, P., and Fung, Y. C. (1971). Slow particulate flow in channel and tube. *J. Appl. Mech. Ser. E* **38**, 721–728.

Tong, P., and Pian, T. H. H. (1967). The convergence of finite element method in solving linear elastic problems. *Int. J. Solids Struct.* **3**, 865–879.

Tong, P., and Pian, T. H. H. (1969). A variational principle and the convergence of a finite element method based on assumed stress distribution. *Int. J. Solids Struct.* **5**, 436–472.

Tong, P., and Pian, T. H. H. (1970). Bounds to the influence coefficients by the assumed stress method. *Int. J. Solids Struct.* **6**, 1429–1432.

Tong, P., and Vawter, D. (1972). An analysis of peristaltic pumping *J. Appl. Mech.* (ASME paper 72-APM-19).

Tong, P., Luk, C. H., and Witmer, E. (1970). "Aeroelastic Study of Structure Composed of Beam Type Structural Elements in Hypersonic Flow," Aeroelastic and Struct. Res. Lab. Rep. No. ASRL-TR-161-1. Massachusetts Institute of Technology, Cambridge, Massachusetts.

Turner, M. J., Clough, R. J., Martin, H. C., and Topp, L. J. (1956). Stiffness and Deflection Analysis of Complex Structures. *J. Aeronaut. Sci.* **23**, No. 9, 805–823.

Visser, W. (1966). A finite element method for the determination of nonstationary temperature distribution and thermal deformation. *Proc. Conf. Matrix Method Struct. Mech., 1st,* 1965 AFFDL-TR-66-80, pp. 925–943.

Washizu, K. (1968). "Variational Methods in Elasticity and Plasticity." Pergamon, Oxford.

Whetstone, W. D., and Yen, C. (1970). "Comparison of Membrane Finite Element Formulations," Rep. No. LMSC/HREC D162553. Lockheed Huntsville Res. Eng. Center, Huntsville, Alabama.

Wilson, E. L., and Nickell, R. E. (1966). Application of the finite element method to heat conduction problem analy. *Nucl. Eng. Des.* **4**, 276–286.

Wolf, J. P. (1971). Programme STRIP pour le calculation des Structures en surface en surface porteuse. *Bull. Tech. Suisse Romande* **97**, **No. 17**, 381–297.

Yamamoto, Y. (1966). "A Formulation of Matrix Displacement Method." Dept. of Aero. and Astro., Massachusetts Institute of Technology, Cambridge.

Zienkiewicz, O. C. (1970). The finite element method: From Intuition to generality. *Appl. Mech. Rev.* **23**, 249–256.

Zienkiewicz, O. C. (1971). "The Finite Element Method in Engineering Sciences." McGraw-Hill, New York.

Zlamal, M. (1968). On the finite element method. *Numer. Math.* **12**, 394–409.

Zudans, Z. (1969). Survey of advanced structural design analysis techniques. *Nucl. Eng. Des.* **10**, 400–440.

The Motion of Bubbles and Drops Through Liquids

J. F. HARPER

Department of Mathematics, Victoria University of Wellington
Wellington, New Zealand

I. Introduction

Chemical engineers, metallurgists, geologists, brewers, and cooks all try to understand processes in which bubbles or drops move through liquids. Until recent years there was not much theoretical analysis to help them, but satisfactory theories now exist for a number of important special cases. The work is interesting mathematically, and also throws light on the physical chemistry of interfaces between moving fluids.

Because interfacial phenomena affect the motion in a number of different ways, there are many cases to consider. We begin in Section II with the simplest. This is a bubble rising in a Newtonian liquid, far enough from boundaries (including other bubbles) to be treated as if isolated. We assume that the viscosity and density of the internal gas are negligible, and that the surface rheology reduces to the single parameter of a constant uniform surface tension. Section III is concerned with drops whose interior viscosity and density are taken into account, but which satisfy the other conditions of Section II.

The most restrictive condition is constancy of the surface tension, for it requires exceptionally pure fluids. Experiments seldom agree with the predictions of Sections II and III for small bubbles or drops unless great care is taken to remove impurities. These adsorb at the surface and lower the surface tension. The flow of the fluid then carries the impurities around the surface, setting up inequalities of concentration and hence of surface tension, which oppose the motion. So much surface-active material may be adsorbed that it forms a layer with a measurable surface elasticity or surface viscosity of its own. If a soluble surfactant diffuses across the interface, it can induce instability of the motion (interfacial turbulence). Only for surfactants so dilute that the motion is stable, and the surface tension and its gradient are the only significant mechanical properties of the interface, do quantitative theories exist. We take the surface tension to depend linearly on the concentrations in the bulk fluids which are proportional to each other, and we assume that the viscosities and the bulk and surface diffusion coefficients are constant. Section IV contains an account of the motion of drops and bubbles in these ideal solutions of surfactants. This is a first step towards a complete theory in rather the same way that the theory of ideal gases is a useful first approximation to the behavior of real gases.

The rheology of nonideal solutions and nonNewtonian liquids lies outside the scope of this review, as do interfacial turbulence, the mutual influences of two or more bubbles or drops, wall effects, and chemical reactions. Information on these topics can be obtained from Brodkey (1967) and Lane and Green (1956) for fluid dynamics, from Adam (1968) and Defay *et al.* (1966) for physical chemistry and thermodynamics of surfaces, and J. T. Davies and Rideal (1963) and Levich (1962) for both physical chemistry and fluid dynamics.

Because moving drops and bubbles are studied by such a wide variety of scientists and engineers, it is useful to indicate where bibliographical information can be found. *Chemical Abstracts* and *Physics Abstracts* are the best general sources of information about current literature, the former in particular having a very broad coverage. *Applied Mechanics Reviews* often has short critical articles on work in the field. Several of the bibliographical reviews published in *Industrial and Engineering Chemistry* were helpful in the preparation of this work (Gal-Or *et al.*, 1969; Gill *et al.*, 1969, 1970, Gomezplata and Regan, 1968; Regan and Gomezplata, 1970; Shedlovsky, 1968; Stahel and Ferrell, 1968; Tavlarides *et al.*, 1970); this journal has now ceased publication and been replaced by *Chemical Technology*. Finally, Gouse (1966) has published an index to over 5000 papers on two-phase gas–liquid flow, up to 1965.

II. A Bubble with Constant Surface Tension Rising Under Gravity

A. Dimensional Analysis

Consider a bubble of gas in an unbounded pure liquid whose surface tension against the gas is σ. We shall assume that σ does not vary around the surface. The bubble will rise steadily if its motion is stable to random small disturbances, and if the time taken to approach very close to the terminal velocity is much less than the time required for the bubble's size to change by a significant fraction of itself. Changes in size may be due to evaporation or condensation, to changes in ambient pressure, or to gases moving in or out of solution (as in a glass of beer). This field of "bubble dynamics" has been reviewed by Plesset (1964), one of the chief contributors to it. Its name is unfortunate, because the theory of rising bubbles of constant size is hardly "bubble statics"!

Suppose that the motion is steady, with a bubble of the same constant volume as a sphere of diameter d (the "equivalent sphere"), rising at speed U in a liquid of density ρ and dynamic viscosity $\eta = \rho\nu$. Let ρ and η be much greater than the density and viscosity of the gas. Then U must be determined by $d = 2a$, ρ, η or ν, σ, and g, the acceleration due to gravity. To make our calculations independent of particular units of measurement and particular fluids, we seek dimensionless products of the above parameters. A number are in common use, but Schmidt (1933) showed that only one can be formed from the given physical properties of the liquid, namely

$$M = g\eta^4/\rho\sigma^3. \tag{2.1}$$

In different liquids M takes values over a very wide range: in highly viscous oils it can exceed 10^5, and in liquid metals it can be less than 10^{-13}. Other dimensionless parameters must depend on U or d or both, and they include the Reynolds number R, drag coefficient C_D, Weber number W, Froude number F, Eötvös number E, and Bond number B, defined by the equations

$$R = Ud/\nu = Ud\rho/\eta, \tag{2.2}$$

$$C_D = \frac{\textit{force on bubble}}{\frac{1}{2}\rho U^2 \cdot \pi a^2} = \frac{\frac{4}{3}\pi \rho g a^3}{\frac{1}{2}\pi \rho U^2 a^2} = \frac{4gd}{3U^2}, \tag{2.3}$$

$$W = \rho U^2 d/\sigma = (4MR^4/3C_D)^{1/3}, \tag{2.4}$$

$$F = U^2/gd = 4/3C_D, \tag{2.5}$$

$$E = B^2 = gd^2\rho/\sigma = \tfrac{3}{4}WC_D. \tag{2.6}$$

The reader is warned that some authors use a in the above definitions in place of d, and others use the bubble's equatorial or polar diameter.

Any independent pair of these dimensionless numbers must determine all the rest in steady flow. It is convenient to choose M, which specifies the fluid, and R which determines the mechanics. Our basic problem is then to calculate the flow pattern and hence C_D as functions of R and M from the equations of motion.

B. Summary of Experimental Results

Haberman and Morton (1953, 1956) and Peebles and Garber (1953) carried out experiments and collected those of previous workers to give graphs of C_D against R for bubbles in a variety of pure liquids. Selected results are shown as solid curves in Fig. 1. (The curve N9 with $M = 1.17 \times 10^{-9}$ is for Bryn's (1933) ethyl alcohol solution in water—Haberman and Morton miscalculated M as 1.17×10^{-8}.) Also in Fig. 1

Fig. 1. Drag coefficient C_D plotted against Reynolds number R for rising bubbles. Experimental curves are drawn solid, theoretical curves dotted. Each curve bears an identifying letter, as in the list following, and also (where relevant) the nearest integer to $-\log_{10} M$. A. Rigid spheres (Perry *et al.*, 1963). B. Tangentially free spheres, $R \ll 1$ (Rybczynski, 1911; Hadamard, 1911). C. Tangentially free spheres, $R \ll 1$, second approximation (Taylor and Acrivos, 1964). D. Tangentially free spheres, $R \gg 1$ (Moore, 1963). E, F, G, H. Tangentially free spheroids (Moore, 1965), M respectively 10^{-6}, 10^{-8}, 10^{-10}, 10^{-12}. I. Spherical caps (Parlange, 1969). J. $M = 1.45 \times 10^{-2}$ (Haberman and Morton, 1953). K. $M = 1.87 \times 10^{-2}$ (Kojima *et al.*, 1968). L. $M = 2.5 \times 10^{-11}$ (Haberman and Morton, 1953). N. $M = 1.17 \times 10^{-9}$ (Haberman and Morton, 1953). P. $M = 6.55 \times 10^{-6}$ (Peebles and Garber, 1953). Q. $M = 2.08 \times 10^{-5}$ (Peebles and Garber, 1953). R. $M = 2.93 \times 10^{-2}$ (Angelino, 1966). S. $M = 3.22 \times 10^{-7}$ (Peebles and Garber, 1953). T. $M = 2.41 \times 10^{-9}$ (Haberman and Morton, 1953). U. $M = 8.9 \times 10^{-11}$ (Haberman and Morton, 1953).

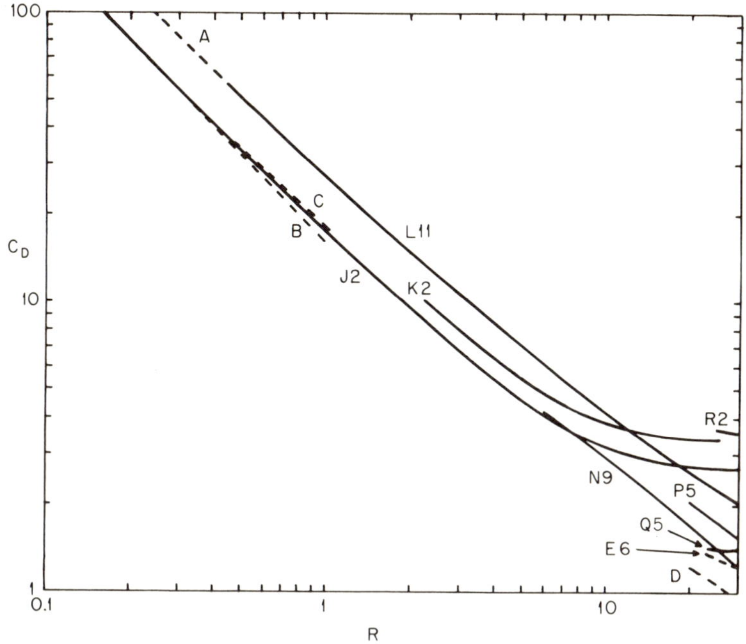
100
10
1
0.1
1
10
C_D
R
A
C
B
L11
J2
K2
R2
N9
P5
Q5
E6
D

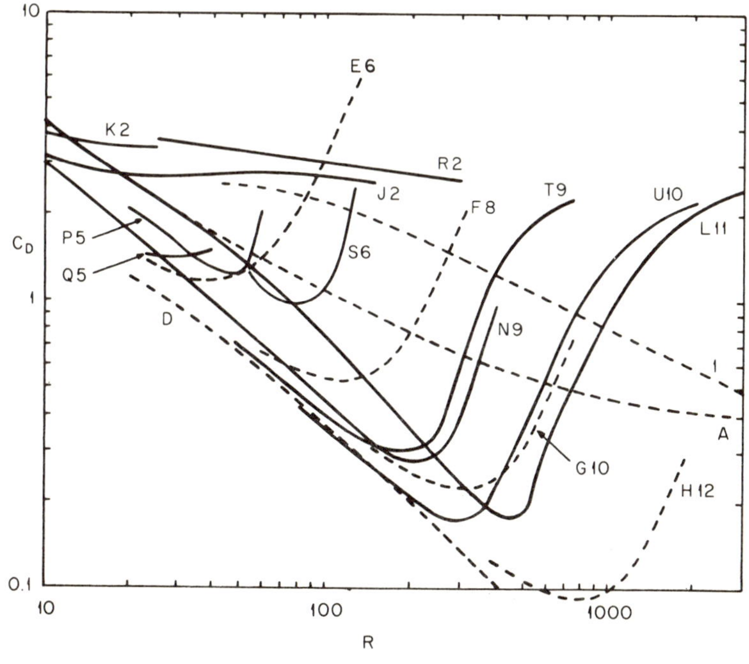
10
1
0.1
10
100
1000
C_D
R
E6
K2
R2
J2
T9
U10
F8
L11
P5
Q5
S6
D
N9
I
A
G10
H12

are the experimental results of Angelino (1966) for $M = 2.93 \times 10^{-2}$ and Kojima *et al.* (1968) for $M = 1.87 \times 10^{-2}$. The dotted curves show $C_D(R)$ for rigid spheres (Perry *et al.*, 1963; Pruppacher *et al.*, 1970) and the theories due to Rybczynski (1911), Hadamard (1911), Taylor and Acrivos (1964), Moore (1963, 1965), and Parlange (1969), which are discussed below. For low values of M (of order 10^{-8} or less) many more results are available but those in Fig. 1 suffice to indicate the general character of all of them. We have used parts only of two of Peebles and Garber's graphs, in a range of M where Haberman and Morton had no liquids, because all their experiments were done in a tube only one inch in diameter. The velocity of the bubbles is appreciably less than that in an "infinite" liquid wherever the equivalent spherical diameter d is greater than about one-fifth of the diameter of the tube. It will be seen that Haberman and Morton's (and Angelino's and Kojima's) graphs tend, at their right-hand ends, towards a fairly constant value of $C_D \doteqdot 2.6$, while Peebles and Garber's values of C_D tend to become proportional to R, because U becomes constant. Many other workers have experimented with bubbles, but the present author has been unable to find any more results with $M > 10^{-6}$, over the range of Reynolds number between 10 and 100, in tubes of sufficiently large diameter.

It is clear that the main features of the experimental graphs are as follows:

1. For each liquid with $M < 10^{-2}$, C_D has a minimum value C_{Dm} which decreases, at a value R_m of R which increases, as M decreases.

2. For values of R less than R_m, C_D rises steadily as R falls, being close to Levich's (1962) asymptotic value of $48/R$ if $R \gg 1$, or to Rybczynski's (1911) and Hadamard's (1911) $16/R$ if $R \ll 1$. Water is a conspicuous exception (curve L11 in Fig. 1) for which C_D is much closer to the value for rigid spheres, as it is in liquids known to be impure. The reason for the anomalous behavior of water is thought (Levich, 1962) to be the presence of trace impurities; see Section IV. In this range of R ($< R_m$) the drag coefficient is not much affected by the value of M.

3. If $R > R_m$, C_D first increases rapidly with R and then levels off at a value of about 2.6, provided that the minimum (C_{Dm}) was lower than that value.

4. If $M > 10^{-2}$ approximately, the minimum of C_D is very shallow or absent. The $C_D(R)$ curves then tend to the Hadamard-Rybczynski asymptote for small R, but level off in the region $2.6 < C_D < 3$. More experimental results are desirable for $R > 10$ and $M > 10^{-6}$; there seems to be a change in the character of the curves between $M = 10^{-5}$ and $M = 10^{-2}$, but no transitional cases are available.

The bubbles rise steadily, and their motion is stable, provided that they are not too large. Hartunian and Sears (1957) collated previous results and also experimented on a number of fluids. Their results are shown in Fig. 2, in which the Weber number W (as defined in Eq. 2.4) is plotted against R for marginally unstable bubbles. The relevant values of $M \times 10^{12}$ are also plotted, as are curves of the dependence of W on R as bubble size varies, for some typical values of M. (As Hartunian and Sears did not tabulate M, values have been found from their references or from the other data in their Table 1. In the process it was discovered that the Bond numbers in that table are all the reciprocals of the values defined by the authors, except for their liquids 13 and 15, for which the table gives ten times the true values.)

For pure liquids, marginal instability occurs at a value of W near 3, provided that M is low enough (less than 10^{-8}) for this to occur when

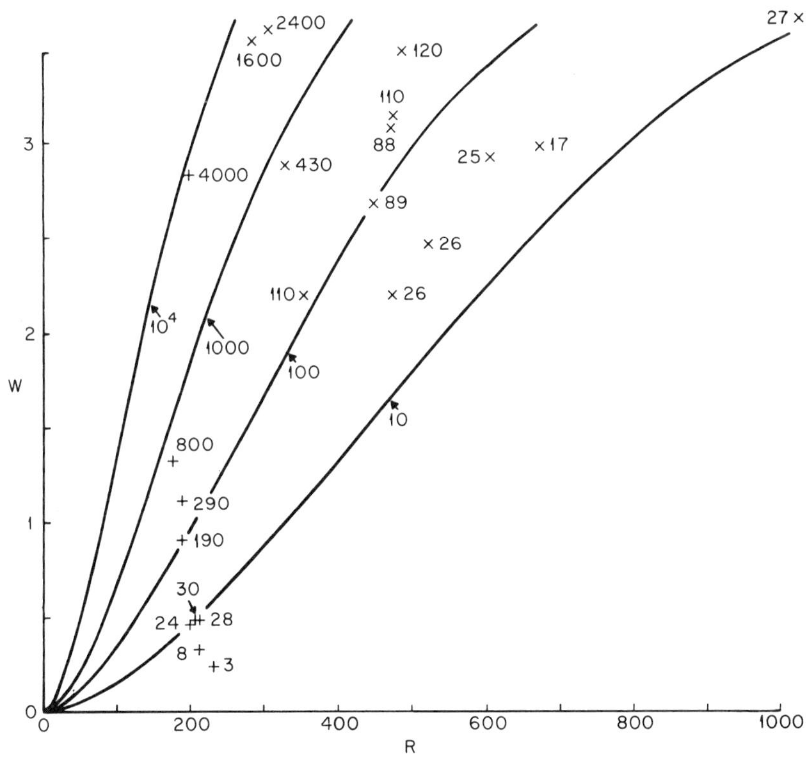

FIG. 2. Critical stability plot, in the (R, W) plane: the curves (based on Fig. 1) show $W(R)$ for four hypothetical liquids with $M = 10^{-8}$, 10^{-9}, 10^{-10}, 10^{-11}. The plotted points indicate marginally unstable bubbles according to Hartunian and Sears (1957), $\times$ for pure liquids, $+$ for contaminated liquids. The associated numbers are values of $10^{12}M$.

$R > 200$. If $M < 10^{-8}$ and the liquid is impure, instability occurs at $R \doteqdot 200$, which requires that $W < 3$. Bubbles are therefore less stable in an impure liquid than in the same liquid pure. No experiments were reported with $M > 10^{-8}$ in which the motion became unstable; in Haberman and Morton's (1953) oil with $M = 1.45 \times 10^{-2}$ bubbles were still stable at $R = 175$, $W = 180$, and Angelino (1966) found no sign of instability up to $R = 300$, $W = 494$ in a mixture of glycols with $M = 2.93 \times 10^{-2}$. When steady vertical motion is unstable, the bubbles rise either in a helical path or in a plane zigzag (see Section II, D).

The other chief property of rising bubbles besides drag and stability is their shape. Sufficiently small bubbles are always spherical, but as W increases towards order unity the bubbles distort, first to oblate spheroids (flattened ellipsoids of revolution) with their short axes of symmetry vertical and their longer equatorial axes horizontal, and ultimately to a spherical cap shape, convex upwards, as C_D becomes constant. Prolate spheroids (rugby football-shaped) are not observed in Newtonian liquids except momentarily during some types of oscillation. We must now turn to the theories which have been advanced to explain this diverse collection of phenomena, taking them in order from left to right across Fig. 1, and then considering Fig. 2.

C. Theory for a Spherical Bubble

1. *Low Reynolds Number*

If the Reynolds number R is very small, viscous stresses which are of order $\eta U/d$ in the neighborhood of the bubble must dominate inertial ones of order $R\eta U/d$, and so the total force exerted by the liquid on the bubble is of order (stress $\times$ area) or ηUd. In steady flow this balances the gravitational upthrust of order ρgd^3, and so U is of order gd^2/ν, and $C_D = 4gd/(3U^2) \propto d^{-3} \propto R^{-1}$. The constant of proportionality depends on the shape of the bubble and the surface conditions. We discuss only shapes close to a sphere; highly distorted bubbles can exist at low Reynolds numbers if M is large enough, but no theory appears to have been published to describe them. The only experiments seem to be those of Pan and Acrivos (1968) and Jones (1965); the undersides of their bubbles became flattened and eventually concave as R increased.

Let us define ψ to be the Stokes (axisymmetric) stream function given in spherical polar coordinates by

$$u_r = -\frac{1}{r^2 \sin\theta}\frac{\partial\psi}{\partial\theta}, \qquad u_\theta = \frac{1}{r\sin\theta}\frac{\partial\psi}{\partial r}, \tag{2.7}$$

where r measures distance from the center of mass of the bubble, making $r = a$ on the surface if it is a sphere, θ is the polar angle measured from the upstream direction, and (u_r, u_θ) are the corresponding velocity components. Then $\psi \sim \frac{1}{2}Ur^2 \sin^2 \theta$ at large distances from the bubble, and we may write the well-known dynamical equation governing Stokes flow as

$$D^4\psi = 0, \tag{2.8}$$

where, with $\cos \theta$ denoted by μ,

$$D^2 = \frac{\partial^2}{\partial r^2} + \frac{1-\mu^2}{r^2} \frac{\partial^2}{\partial \mu^2}. \tag{2.9}$$

This equation is known to give a good approximation to the real flow for small R provided that $r \ll a/R$ (see, for example, Happel and Brenner, 1965).

The general solution of (2.8) with the correct limiting form for $r \gg a$ is given by

$$\frac{\psi}{Ua^2} = \left(\frac{r^2}{a^2} - \alpha_2 \frac{r}{a} + \beta_2 \frac{a}{r}\right)\mathscr{I}_2(\mu) - \sum_{n=3}^{\infty}\left(\alpha_n \frac{a^{n-3}}{r^{n-3}} - \beta_n \frac{a^{n-1}}{r^{n-1}}\right)\mathscr{I}_n(\mu), \tag{2.10}$$

where $\mathscr{I}_n(\mu)$ is defined in terms of the Legendre polynomial $P_{n-1}(\mu)$ for $n = 2, 3, 4, \ldots$ by

$$\mathscr{I}_n(\mu) = \int_\mu^1 P_{n-1}(x)\, dx = \frac{1-\mu^2}{n(n-1)} \frac{dP_{n-1}(\mu)}{d\mu}. \tag{2.10}$$

For sufficiently high surface tension (or low M) the bubble may be assumed with very little error to be a sphere, on the surface of which we may put $\psi = 0$, so that $\beta_2 = \alpha_2 - 1$ and $\beta_n = \alpha_n$ for $n \geq 3$ in (2.10). Then the drag coefficient may be calculated by integrating the streamwise component of stress over the surface of the sphere, or by finding the momentum flux at a distance (Happel and Brenner, 1965), as

$$C_\text{D} = 16\alpha_2/R. \tag{2.12}$$

To render the solution unique and find α_2 we need another surface condition. If, for example, the sphere were rigid, making $u_\theta = 0$ at $r = a$, the use of (2.7) and (2.8) would immediately give Stokes's results $\alpha_n = \beta_n = 0$ for $n \geq 3$, $\alpha_2 = 3/2$, and $C_\text{D} = 24/R$. In our case of constant surface tension the surface shear stress component $p_{r\theta}$ vanishes (see Levich, 1962; Landau and Lifshitz, 1959), and because

$$p_{r\theta} = \frac{\eta a}{\sin \theta} \frac{\partial}{\partial r}\left(\frac{1}{r^2} \frac{\partial \psi}{\partial r}\right)$$

on the surface, where $u_r = 0$, we have for all μ in $-1 \leq \mu \leq 1$,

$$p_{r\theta} = \frac{\eta U}{a \sin \theta} [6(\alpha_2 - 1)\mathscr{I}_2(\mu) + \sum_{n=3}^{\infty} (4n - 2)\alpha_n \mathscr{I}_n(\mu)], \tag{2.13}$$

which leads to the Rybczynski-Hadamard results $\alpha_n = \beta_n = 0$ for $n \geq 3$, $\alpha_2 = 1$, $C_D = 16/R$, and $u_\theta = \frac{1}{2}U \sin \theta$ at $r = a$.

The method used above is exactly similar for free and rigid surfaces, only the boundary conditions and therefore the numerical values being different. The same is true for higher approximations in R: one uses matched Stokes and Oseen asymptotic expansions in $r \ll Ra$, $r \gg a$ respectively, in the manner pioneered by Lagerstrom and Cole (1955) and Proudman and Pearson (1957). Taylor and Acrivos (1964) carried the work through, for drops as well as bubbles, with tangentially stress-free surfaces. They found higher approximations to the drag, the next for a bubble being $C_D = (16/R) + 2$, and showed how to determine the first-order perturbations of shape from a sphere. If the surface of the bubble is at

$$r = a[1 + \zeta(\mu)], \qquad \max|\zeta| \ll 1, \tag{2.14}$$

the conditions that a be the radius of the equivalent sphere and that the origin be the center of mass are, to first order,

$$\int_{-1}^{1} \zeta(\mu)\, d\mu = \int_{-1}^{1} \mu\zeta(\mu)\, d\mu = 0. \tag{2.15}$$

We also require that the difference between the normal components of stress just inside and just outside the surface be the surface tension times the sum of the principal curvatures. Using this condition, Taylor and Acrivos found for a gas bubble that

$$\zeta = -\tfrac{5}{48} W P_2(\mu) = \tfrac{5}{96} W(1 - 3\cos^2\theta), \tag{2.16}$$

to the first order (we recall that $W = \rho U^2 d/\sigma$). The bubble is thus an oblate spheroid with its short axis vertical, the ratio χ of longest to shortest diameter being given by

$$\chi = 1 + \tfrac{5}{32} W. \tag{2.17}$$

We shall not here pursue higher approximations to the shape or the drag: Taylor and Acrivos give some, which predict flattened undersides of rising bubbles, but their neglected terms are less important only if $W \ll 1$, $R \ll 1$, and $\rho\nu^2/a\sigma \ll 1$, i.e. $R \ll M$. The last inequality is frequently too restrictive: it would be helpful to have also the next approximation for $W \ll 1$, $R \ll 1$ and $R \gg M$. Pan and Acrivos (1968) have, however, published some experimental results in an oil with $M = 171$ which agreed with the theory for $R \ll M$.

The surprising point about Eq. (2.17) is that $\chi - 1$ is proportional to W, a ratio of inertial to surface tension stress. One would have expected $\eta U/\sigma$ instead, because viscous stresses dominate inertial ones in determining the motion, but it happens that terms of order $\eta U/\sigma$ cancel in the derivation of (2.16). They must, of course, for a shape symmetrical about the equatorial plane, like the spheroid of Eq. (2.16): reversing U would not alter the shape, and so the distortion must be an even function of U. As a result, a spherical bubble is an exact solution in Stokes flow with inertial terms neglected. We observe finally that $\chi - 1$ is a rapidly increasing function of R: if $C_D = 16/R$, $W = (32/5)(\chi - 1) = (MR^5/12)^{1/3}$, by Eq. (2.4).

2. *High Reynolds Numbers*

For $R \gg 1$, the theories of motion past solid spheres and tangentially stress-free bubbles are quite different. It is easy to see why this must be so. In either case vorticity must be generated at the surface because irrotational flow does not satisfy all the boundary conditions. The vorticity remains within a boundary layer of thickness $\delta = O(aR^{-1/2})$, for it is convected around the surface in a time t of order a/U, during which viscosity can diffuse it away to a distance δ if $\delta^2 = O(\nu t) = O(a^2/R)$. But for a solid sphere the fluid velocity must change by $O(U)$ across the layer, because it vanishes on the surface, whereas for a gas bubble the normal derivative of velocity must change by $O(U/a)$ in order that the shear stress be zero. That implies that the velocity itself changes by $O(U\delta/a) = O(UR^{-1/2}) = o(U)$, as was first pointed out by Levich (1949).

In the boundary layer on the bubble, therefore, the fluid velocity is only slightly perturbed from that of the irrotational flow, and velocity derivatives are of the same order as in the irrotational flow. Then the viscous dissipation integral has the same value as in the irrotational flow, to the first order, because the total volume of the boundary layer, of order $a^2\delta$, is much less than the volume, of order a^3, of the region in which the velocity derivatives are of order U/a. The volume of the wake is not small, but the velocity derivatives in it are, and it contributes to the dissipation only in higher order terms (Moore, 1963). By evaluating the dissipation in irrotational flow past a sphere, Levich (1949) obtained

$$C_D = 48/R, \tag{2.18}$$

as a first approximation to the drag coefficient of a bubble at high R, as did Ackeret (1952) and Chan and Prince (1965) later but independently. The bubble must be spherical if the surface tension pressure, of order σ/a, is much greater than the dynamic pressure of order ρU^2, i.e. if $W \ll 1$.

Further developments have been remarkable for the number of errors perpetrated: most writers on this subject seem to have made at least one in published work. Let us begin by considering the velocity distribution in the boundary layer, assuming the bubble to be spherical and using the same spherical polar coordinate system as at low Reynolds numbers. Let the θ velocity component $u_\theta = \bar{u}_\theta + u_\theta'$, where $\bar{u}_\theta$ denotes the velocity of the irrotational flow past the sphere and u_θ' the perturbation due to viscosity. Then

$$\bar{u}_\theta = U \sin\theta\left(1 + \frac{a^3}{2r^3}\right), \tag{2.19}$$

as is well known. To find u_θ' it is convenient to start by defining a "circulation density" Ω in terms of the azimuthal vorticity component ω

$$\Omega = \frac{\omega}{r\sin\theta} = \frac{D^2\psi}{r^2\sin^2\theta} = \frac{1}{r^2\sin^2\theta}\left[\frac{\partial^2\psi}{\partial r^2} + \frac{\sin\theta}{r^2}\frac{\partial}{\partial\theta}\left(\frac{1}{\sin\theta}\frac{\partial\psi}{\partial\theta}\right)\right]. \tag{2.20}$$

The reason for the name "circulation density" (for which I am indebted to Prof. J. E. Ffowcs Williams) is that Ω is 2π times the circulation round an infinitesimal vortex tube, divided by the volume of the tube.

The form of the vorticity equation in terms of Ω, for time-dependent axially symmetric flow in general orthogonal curvilinear coordinates (α, β, γ) is frequently useful, and does not appear in the usual reference books. If (α, β, γ) are a right-handed set of coordinates in which γ is the azimuthal angle, and the element of distance ds is given by

$$(ds)^2 = h_1^2(d\alpha)^2 + h_2^2(d\beta)^2 + m^2(d\gamma)^2, \tag{2.21}$$

where m is distance from the axis of symmetry, and the velocity components u_α, u_β are given by

$$u_\alpha = -\frac{1}{mh_2}\frac{\partial\psi}{\partial\beta}, \qquad u_\beta = \frac{1}{mh_1}\frac{\partial\psi}{\partial\alpha}, \tag{2.22}$$

then the vorticity equation can be written as

$$\frac{\partial\Omega}{\partial t} + \frac{1}{mh_1h_2}\frac{\partial(\psi, \Omega)}{\partial(\alpha, \beta)} = \nu\left(\nabla^2\Omega + \frac{2}{m}\nabla m \cdot \nabla\Omega\right), \tag{2.23}$$

after a simple transformation from the form given by Goldstein (1938), where the circulation density Ω is given by

$$\Omega = \frac{1}{m^2}D^2\psi = \frac{1}{mh_1h_2}\left[\frac{\partial}{\partial\alpha}\left(\frac{h_2}{mh_1}\frac{\partial\psi}{\partial\alpha}\right) + \frac{\partial}{\partial\beta}\left(\frac{h_1}{mh_2}\frac{\partial\psi}{\partial\beta}\right)\right]. \tag{2.24}$$

In our thin small-perturbation boundary layer in steady flow, with α, β respectively the stream function $\bar{\psi}$ and velocity potential $\bar{\varphi}$ of the irrotational first approximation, and $\bar{v}$ the speed of the fluid in that approximation, $h_1 = 1/(m\bar{v})$, $h_2 = 1/\bar{v}$. To a sufficient approximation (2.23) is then

$$\frac{1}{m^2}\frac{\partial\Omega}{\partial\bar{\varphi}} = \nu\frac{\partial^2\Omega}{\partial\bar{\psi}^2}. \tag{2.25}$$

As variations of m across the boundary layer may be neglected, m may be taken to be the function of $\bar{\varphi}$ alone with which it coincides on the bubble surface $\bar{\psi} = 0$. Equation (2.25) then reduces to

$$\partial\Omega/\partial X = \nu\ \partial^2\Omega/\partial\bar{\psi}^2, \tag{2.26}$$

if $X = \int_0^{\bar{\varphi}} [m(\varphi)]^2\, d\varphi$. To a boundary layer approximation the perturbation tangential velocity $u_\alpha' = u_{\bar{\phi}}'$ is given by

$$u_\alpha'(X, \bar{\psi}) = -\frac{1}{\bar{v}}\int_{\bar{\psi}}^{\infty} \Omega(X, \psi)\, d\psi, \tag{2.27}$$

because u_α' and Ω both tend to zero at the outer limit of the boundary layer, to a first approximation in which we neglect the secondary flow due to displacement thickness (Lighthill, 1958). Hence $\bar{v}u_\alpha'$ also obeys the same one-dimensional diffusion equation (2.26) as Ω. Note that if $\bar{\varphi}$ on the bubble surface is proportional to distance s measured parallel to the axis of symmetry (and it is for spheres and spheroids), then X is proportional to the volume of the bubble upstream from a plane of constant s.

The subsequent working is simplified if we choose dimensionless variables $x \propto X$, $z \propto \bar{\psi}$, and $f(x, z) \propto \bar{v}u_\alpha$ such that $0 \leq x \leq 1$ on the surface of the bubble, $f(x, z)$ is of order unity in the boundary layer, and Eq. (2.26) becomes

$$4\ \partial f/\partial x = \partial^2 f/\partial z^2. \tag{2.28}$$

These requirements are met for a spherical bubble by putting

$$x = \tfrac{1}{4}(2 - 3\mu + \mu^3) = \tfrac{1}{4}(1 - \mu)^2(2 + \mu), \tag{2.29}$$

where $\mu = \cos\theta$ as before,

$$\begin{aligned} z &= 3R^{1/2}(r - a)\sin^2\theta/8a \\ &= \bar{\psi}R^{1/2}/4Ua^2, \end{aligned} \tag{2.30}$$

and

$$f(x, z) = R^{1/2}u_\alpha' \sin\theta/U. \tag{2.31}$$

The boundary conditions on f are, firstly, the initial condition that the fluid entering the boundary layer near the front stagnation point has no velocity perturbations, i.e.

$$f(0, z) = 0 \quad \text{for} \quad z > 0, \tag{2.32}$$

and secondly, the surface shear stress condition which reduces to

$$\frac{\partial f}{\partial z}(x, 0) = 8 \quad \text{for} \quad 0 < x < 1. \tag{2.33}$$

The solution of Eq. (2.28) with these conditions is immediately found to be $\partial f/\partial z = 8 \operatorname{erfc}(z/x^{1/2})$, or, on integrating,

$$f(x, z) = -8x^{1/2} \operatorname{ierfc}(z/x^{1/2}) = -8x^{1/2} \int_{z/x^{1/2}}^{\infty} \operatorname{erfc}(t)\, dt. \tag{2.34}$$

This result was first found correctly by Moore (1963) in a different notation from that used here; Levich (1962) and Chao (1962) used erroneous forms of the continuity and stress equations, respectively.

The perturbation pressure is extraordinarily difficult to find. Chao took the difference between the actual pressure in the fluid and the pressure in the irrotational flow to be $o(\rho U^2 R^{-1})$; Moore (1963) showed that it was $O(\rho U^2 R^{-1})$, but he has since pointed out (personal communication) that he neglected the secondary flow due to displacement thickness, which contributes additional terms to his paper's Eqs. (2.22), (2.36), and (2.37) of the same order as the ones which he did take into account. Moore's drag calculation does not depend on those equations and is correct to the order given by him, but Chao's is vitiated by his error in determining the pressure. Before describing Moore's method for the drag, we must look into the behavior of $u_\theta' \doteqdot u_\alpha'$ near the rear stagnation point, as it has been widely misunderstood. It is clear from our Eq. (2.34) that on the surface $z = 0$, near the rear stagnation point $x = 1$, $f(x, z) \doteqdot -8\pi^{-1/2}$, or $u_\theta' \doteqdot -8UR^{-1/2}\,\pi^{1/2} \operatorname{cosec} \theta$, and $\bar{u}_\theta = (3/2)U \sin \theta$. It would therefore seem that the resultant θ-velocity $u_\theta = \bar{u}_\theta + u_\theta'$ must reverse and the boundary layer must separate, where $\sin^2 \theta = 16/(3\pi^{1/2}\, R^{1/2})$, i.e. $\pi - \theta = O(R^{1/4})$. But Moore (1963) showed that if $\pi - \theta = O(R^{-1/6})$ the above theory giving u_θ' is not valid and must be replaced by a different one, which predicts $u_\theta' \ll \bar{u}_\theta$ and hence no boundary layer separation.

Moore's theory for the rear stagnation region proceeds as follows. One writes down the full equations of motion, and evaluates the order of magnitude of each term with the aid of the boundary layer solution (2.34). The result is that if $\varphi = \pi - \theta$ is in the range $R^{-1/8} \ll \varphi \ll 1$ then the terms retained in the approximation (2.28) do dominate those neglected, making the approximation still consistent, and the inertial terms also dominate the viscous ones. (Some of the retained terms are therefore negligible, but that is merely harmless additional complication in the theory: for consistency one requires only that all neglected terms be negligible.) In other words, if $R^{-1/8} \ll \varphi \ll 1$ the assumption of a thin boundary layer still holds and the perturbation velocity is still much smaller than the irrotational velocity at each point, but viscous diffusion of circulation density may be ignored, as if the fluid were inviscid. The physical explanation for this inviscid behavior is simply that streamlines must be further apart near stagnation points than in the flow generally, and so space derivatives of circulation density are reduced, which impedes its diffusion and leaves what is already there to be carried along the streamlines.

This reason for the flow to be effectively inviscid where $R^{-1/8} \ll \varphi \ll 1$ still holds where $\varphi = O(R^{-1/8})$ and the fluid which was formerly in the boundary layer around the surface turns the corner and passes down the wake. Streamlines whose distances from most of the bubble surface are of order $aR^{-1/2}$ pass the stagnation points at distances of order $aR^{-1/6}$ and lie at distances of order $aR^{-1/4} \gg aR^{-1/2}$ from the center-line of the wake, and so diffusion of circulation density across them is a much slower process in the rear stagnation region and wake than in the viscous part of the boundary layer around the surface. This argument could fail if the "inviscid" flow did not obey the viscous boundary condition of zero shear stress on the surface, but (unusually for an inviscid flow) that condition turns out to be satisfied near the rear stagnation point, to leading order. It is not satisfied near the front stagnation point, which explains why the viscous layer begins there, rather than at some small distance downstream.

Because the "inviscid" region has $\varphi \ll 1$, we may use cylindrical polar coordinates (m, s) to describe it, where m is the perpendicular distance from the axis of symmetry, s is the distance downstream from the rear stagnation point measured parallel to the axis, and the bubble surface approximates to the plane $s = 0$ if φ is small. Then, if $\bar{\psi}$ and ψ' are the stream functions of the irrotational flow and the perturbation to it, the circulation density $\Omega = \omega/m$ obeys

$$\Omega = \frac{1}{m^2}\left(\frac{\partial^2\psi'}{\partial m^2} - \frac{1}{m}\frac{\partial\psi'}{\partial m} + \frac{\partial^2\psi'}{\partial s^2}\right) = B(\bar{\psi} + \psi'), \tag{2.35}$$

in the inviscid region, where B is a function to be determined from the requirement of matching to the upstream flow at $a \gg m \gg aR^{-1/8}$. There, $\bar{\psi} \gg \psi'$ and

$$\Omega \doteqdot \frac{1}{a\sin\theta}\frac{\partial u_\theta'}{\partial r} \doteqdot \frac{3U}{a}\operatorname{erfc} z, \tag{2.36}$$

and hence, in the stagnation region,

$$B(\bar{\psi}) \doteqdot \frac{3U}{a^2} \operatorname{erfc}\left(\frac{\bar{\psi}R^{1/2}}{4Ua^2}\right) \doteqdot \frac{3U}{a^2} \operatorname{erfc}\left(\frac{3R^{1/2}m^2s}{8a^3}\right) = b(m^2s), \quad (2.37)$$

say. One can now find a self-consistent approximation to the solution of (2.35) and (2.37). Let us define dimensionless stagnation region variables m_1, s_1, to be $mR^{1/6}/a$, $sR^{1/6}/a$, respectively, and assume that $\psi' \ll \bar{\psi}$ in the region where m_1 and s_1 are of finite order. Then

$$\frac{\partial^2\psi'}{\partial {m_1}^2} - \frac{1}{m_1}\frac{\partial\psi'}{\partial m_1} + \frac{\partial^2\psi'}{\partial {s_1}^2} = 3Ua^2R^{-2/3} \operatorname{erfc}(\tfrac{3}{8}{m_1}^2 s_1), \quad (2.38)$$

and $\psi' = 0$ on $s_1 = 0$ $m_1 = 0$. The chief quantity of interest is $u_\theta' = -u_m' = (1/m)(\partial\psi'/\partial s)$ evaluated on the bubble surface $s = 0$. From (2.38) that is given in terms of m_1 by

$$u_\theta' = -3UR^{-1/3} \int_0^\infty \int_0^\infty \int_0^\infty s^2ke^{-kz}J_1(ks)J_1(km_1)\operatorname{erfc}(\tfrac{3}{8}s^2z)\, dz\, ds\, dk. \quad (2.39)$$

In the limit $R \to \infty$ this equation holds for all m_1. For large m_1, u_θ' can be shown (Harper and Moore, 1968) to be

$$u_\theta' \sim -\frac{1}{m}\int_0^\infty b(t)\, dt \sim -\frac{8UR^{-1/3}}{\pi^{1/2}m_1} \sim -\frac{8UR^{-1/2}}{\pi^{1/2}\varphi}, \quad (2.40)$$

where $b(t)$ is the function defined in (2.37). This form agrees with the boundary layer solution for small φ, so that the present theory, derived for finite $m_1 = \varphi R^{1/6}$, is also valid in the region of overlap with the boundary layer, $R^{-1/8} \ll \varphi \ll 1$ or $R^{1/24} \ll m_1 \ll R^{1/6}$. However, $|u_\theta'|$ does not go on increasing at the rate indicated by (2.40) as m_1 decreases to order unity, but remains $O(UR^{-1/3})$ in the stagnation region where $\bar{u}_\theta = O(UR^{-1/6})$. Near the rear stagnation point where m_1 is small, $J_1(km_1) \sim \frac{1}{2}km_1$ in (2.39), and so

$$u_\theta' \sim -0.5749m \int_0^\infty b(t^3)\, dt \sim -1.83UR^{-1/3}m_1 \sim$$

$$-1.83UR^{-1/6}\varphi \sim -1.22R^{-1/6}\bar{u}_\theta, \quad (2.41)$$

from Harper and Moore's (1968) equations (7.4), (7.9) with the sign error corrected, and (7.10).

We now see that $u_\theta' \ll \bar{u}_\theta$ everywhere around the bubble if the Reynolds number is high. The ratio $|u_\theta'/u_\theta|$ is of order $R^{-1/2}$ over most of the surface ($\varphi > o(1)$), of order $R^{-1/2}/\varphi^2$ if $\varphi \gg R^{-1/6}$, and of order $R^{-1/6}$ if $\varphi = O(R^{-1/6})$. Therefore the boundary layer does not separate, $\bar{\psi} \gg \psi'$ even in the stagnation region, and so our approximations are consistent. The point has been made here at some length, because several authors (Levich, 1962; Chao, 1962; Winnikow and Chao, 1966; Taunton and Lightfoot, 1969) have used the simple viscous boundary layer analysis well beyond its domain of validity and deduced incorrect results about separation.

None of this theory says, of course, that separation and the associated back-eddies in the wake do not occur in contaminated fluids or at Reynolds numbers of order 10: we have given only an asymptotic analysis for spherical bubbles in pure fluids at high Reynolds numbers. The theory predicts, for example, that there is a neighborhood of the rear stagnation point where $|u_\theta'| > \bar{u}_\theta$ if $R < 3.3$ and one where $|u_\theta'| > \frac{1}{2}\bar{u}_\theta$ if $R < 210$. Even $\frac{1}{2}u_\theta'$ is not a very small perturbation, but the region in which the perturbations are this large is of limited size; this presumably explains the good agreement sometimes obtained between theoretical and experimental values of C_D (see below).

Having determined the nature of the velocity perturbations, we can now evaluate the viscous drag on the bubble. One makes most efficient use of the available information by choosing a method which gives the drag correctly to first order without any knowledge of the details of the boundary layers. It is then possible to find a second approximation to the drag from the first-order perturbations to the irrotational flow. We therefore reject the obvious methods of integrating the momentum defeat in the wake [but see Moore's (1963) proof that it gives the first-order drag correctly], or the normal stress on the surface, in favor of an energy argument. In steady flow, the rate of working of gravity on the bubble, which is the drag force times the velocity U, is equal to the rate of working against the surface stress plus the rate of viscous dissipation of energy throughout the fluid. Moore's (1963) final result from this method is

$$C_D = \frac{48}{R}\left[1 - \frac{2.211}{R^{1/2}} + O(R^{-5/6})\right], \tag{2.42}$$

where the leading term represents the irrotational dissipation, the next is due to the boundary layer and wake, and the error term comes principally from neglecting viscous dissipation and surface stress in the rear stagnation region. It is of interest to observe that a necessary condition for the analysis to hold is therefore $R^{-5/6} \ll R^{-1/2}$, or $R^{-1/3} \ll 1$, but the best sufficient condition we have is $R^{-1/6} \ll 1$.

Equation (2.42) is plotted in Fig. 1 as curve D. It gives a good description of the experimental facts for bubbles which satisfy its conditions of validity, i.e. high R (over 50 appears to suffice), low enough M for the bubble to be almost spherical for some range of R over 50 ($M < 10^{-8}$ will do), and fluids pure enough for σ to be effectively constant. But Fig. 1 also shows that C_D does not continue to decrease in the way indicated by (2.42) for any fluid, but begins after a certain value of R to rise steeply, especially if M is very small. We now turn to the theory of that effect.

D. Spheroidal Bubbles at High Reynolds Numbers

So far, the theory for high Reynolds numbers has described only spherical bubbles, but in practice the bubbles are frequently of very different shapes. It is not hard to see why: let the equation of the surface be, as in Section II, C, 1,

$$r = a[1 + \zeta(\mu)] = a[1 + \zeta(\cos\theta)], \tag{2.14}$$

and suppose that $\max|\zeta| \ll 1$, so that

$$\int_{-1}^{1} \zeta(\mu)\, d\mu = \int_{-1}^{1} \mu\zeta(\mu)\, d\mu = 0. \tag{2.15}$$

Suppose also that the flow has the general nature of an irrotational motion slightly perturbed by boundary layers, as above, and then the pressure is nearly (minus) the normal stress component at the surface. The error is of order $\rho U^2/R$ (Moore, 1959), and so the surface condition is

$$p_1 = p_0 + \sigma(\kappa_1 + \kappa_2) + O(\rho U^2/R), \tag{2.43}$$

where p_1, p_0 are the pressures just inside and outside the surface and κ_1, κ_2 are its principal curvatures. Bernoulli's theorem and elementary differential geometry then give

$$\begin{aligned} -a(\kappa_1 + \kappa_2) &= -2 + (1 - \mu^2)\zeta'' - 2\mu\zeta' + 2\zeta + O(\zeta^2) \\ &= -\tfrac{9}{16}W + O(\zeta^2) + O(WC_D) + O(W/R) + \text{constant}, \end{aligned} \tag{2.44}$$

where W is the Weber number $2\rho U^2 a/\sigma$, the term in WC_D comes from hydrostatic pressure differences around the bubble, and the constant term appears because the pressure p_1 inside the bubble is, so far, unspecified. The solution to (2.44) with the conditions (2.15) is

$$\begin{aligned} \zeta &= -\tfrac{3}{32}WP_2(\mu) + O(W^2) + O(WC_D) + O(W/R) \\ &= -\tfrac{3}{64}W(3\mu^2 - 1) + O(W^2) + O(WC_D) + O(W/R), \end{aligned} \tag{2.45}$$

which is a useful approximation if $W \ll 1$, $R \gg 1$ and $C_D \ll 1$. It gives an axis ratio χ of the bubble equal to $1 + 9W/64$, to leading order. The bubble's distortion is initially to the same oblate spheroidal shape at high or low Reynolds numbers, and the coefficients of W differ only by 10%, χ being $1 + 5W/32$ for $R \ll 1$.

In spite of the limitations imposed in the above theory, oblate spheroids are found to be fair approximations to the true shapes of bubbles for quite large values of W. Accordingly Moore (1965) developed the theory for spheroids analogous to his previous one (Moore, 1963) for spheres. As in Section II.3 above, one calculates the viscous boundary-layer corrections to the irrotational axially symmetric flow, and obtains the first two terms for the drag as

$$C_{\mathrm{D}} = \frac{48}{R} G(\chi)\left(1 + \frac{H(\chi)}{R^{1/2}}\right) \tag{2.46}$$

where $G(\chi)$ and $H(\chi)$ are functions given in Table 1. $G(\chi)$ represents the

TABLE 1

MOORE'S FUNCTIONS FOR DISTORTED BUBBLES[a]

χ	$W(\chi)$	$G(\chi)$	$H(\chi)$	χ	$W(\chi)$	$G(\chi)$	$H(\chi)$
1.0	0.000	1.000	−2.211	2.6	3.278	4.278	+1.499
1.1	0.624	1.137	−2.129	2.7	3.325	4.565	1.884
1.2	1.108	1.283	−2.025	2.8	3.368	4.862	2.286
1.3	1.492	1.437	−1.899	2.9	3.406	5.169	2.684
1.4	1.802	1.600	−1.751	3.0	3.441	5.487	3.112
1.5	2.056	1.772	−1.583	3.1	3.473	5.816	3.555
1.6	2.268	1.952	−1.394	3.2	3.501	6.155	4.013
1.7	2.446	2.142	−1.186	3.3	3.527	6.505	4.484
1.8	2.597	2.341	−0.959	3.4	3.550	6.866	4.971
1.9	2.727	2.549	−0.714	3.5	3.572	7.237	5.472
2.0	2.839	2.767	−0.450	3.6	3.591	7.620	5.987
2.1	2.937	2.994	−0.168	3.7	3.608	8.013	6.517
2.2	3.022	3.231	+0.131	3.8	3.624	8.418	7.061
2.3	3.098	3.478	+0.448	3.9	3.369	8.834	7.618
2.4	3.165	3.735	+0.781	4.0	3.652	9.261	+8.189
2.5	3.224	4.001	+1.131				

[a] The values of $W(\chi)$ and $G(\chi)$ have been calculated from Moore's (1965) formulas; the values of $H(\chi)$ have been copied from his paper, by permission of the Cambridge University Press.

irrotational dissipation rate E, which can for a general three-dimensional body (Harper, 1970, 1971) be put in either of the forms

$$E = -\eta \iint_S \frac{\partial v^2}{\partial n}\, dS = 2\eta \iint_S \left(u_\beta{}^2 h_\gamma \frac{\partial h_\beta}{\partial n} + u_\gamma{}^2 h_\beta \frac{\partial h_\gamma}{\partial n}\right) d\beta\, d\gamma. \tag{2.47}$$

Here (α, β, γ) are orthogonal curvilinear coordinates in which the bubble S has the equation $\alpha = \text{constant}$, $\partial/\partial n = (1/h_\alpha)(\partial/\partial\alpha)$ denotes differentiation in the direction normal to S and into the fluid, the element of length ds is given by $(ds)^2 = h_\alpha{}^2(d\alpha)^2 + h_\beta{}^2(d\beta)^2 + h_\gamma{}^2(d\gamma)^2$, v is the speed of the fluid, and u_β, u_γ the components of velocity in the β and γ directions. The first form for E in (2.47) is classical (see Lamb, 1932) but has the drawback of requiring velocity gradients. The second requires gradients of metric components instead, which are usually easier to find.

A spheroid will not satisfy the pressure condition (2.43) everywhere around the surface, and so some approximation has to be made. Two methods spring to mind: either equating as many terms as possible of a series expansion near the front stagnation point (R. M. Davies and Taylor, 1950; Saffman, 1956), or satisfying the condition only at the equator and poles (Hartunian and Sears, 1957; Moore, 1959, 1965). The latter method gives results which agree better with experiments, and the values of $W(\chi)$ predicted by it are shown in Table 1. The data in that table, together with the identity

$$C_D = \tfrac{4}{3}MR^4W^{-3}, \tag{2.48}$$

allow one to calculate $C_D(R, M)$. Curves for $M = 10^{-6}$, 10^{-8}, 10^{-10}, and 10^{-12} will be found in Fig. 1. Each curve extends from $\chi = 0.1$, where C_D is close to the value for spherical bubbles (Moore, 1963), to $\chi = 4.0$, where W has become nearly constant, making $C_D \propto R^4$ approximately. The limiting value of W from Moore's theory is 3.745. Although the theory is then beyond its range of validity (see the next paragraph), experiments indicate that W is not far from that value when it becomes a slowly varying function of χ. The shape of the bubble is then very sensitive to small changes in W, which might be caused by small currents in the surrounding fluid. It is therefore not surprising that bubbles in pure low-M liquids become unstable for values of W greater than about 3.

The type of motion which appears when a steady rise of bubbles in a straight line becomes unstable is either steady motion relative to the bubble up a helix with a vertical axis, or else zigzagging in a vertical plane on either side of a vertical line. In spite of many careful experiments, reviewed by Saffman (1956) and Hartunian and Sears (1957), who also performed their own, there is no agreement as to the conditions which decide between the two modes of instability. The above-named authors also gave approximate theories to account for the motion. Saffman used a series expansion of the irrotational-flow pressure condition near the top stagnation point, and showed that steady motion became unstable to zigzags for axis ratios $\chi > 1.2$, and that spiraling was possible for $1.2 < \chi < 2.2$, although it would only occur for motions suitably started. Experimentally, these values of χ are too low. Hartunian and Sears obtained a more realistic result for the onset of zigzagging by satisfying the pressure condition at the poles and equator, namely $\chi > 2.2$, but they did not find any criterion for spiraling. Both methods gave exponentially increasing sideways displacements instead of the oscillatory behavior actually seen in zigzagging bubbles. Saffman suggested that the oscillations might be akin to those seen in the wakes of solid bodies at Reynolds numbers of a few hundred.

A glance at Fig. 1 will show that the rising theoretical curves for $C_D(R, M)$ are fairly close to the experimental ones provided that $C_D < 1$. This limitation shows the importance of the hydrostatic error term in Eq. (2.45), but it is less clear why Moore's (1965) theory works as well as it does for distorted bubbles. The predicted perturbations are not small. To see this, let us assume that the bubble is a very flattened spheroid and keep only the leading terms for large χ in Moore's analysis. His equation analogous to (2.32) can be written

$$u_\theta' = \frac{-2\sqrt{2}U(1+\chi^2\cos^2\theta)^{1/2}}{\sqrt{3}\pi\chi^{1/6}R^{1/2}\sin\theta}\int_0^X \frac{S(\tau)}{(X-\tau)^{1/2}}\exp\left\{-\frac{Y^2}{4(X-\tau)}\right\}d\tau, \tag{2.49}$$

where θ is the eccentric angle of a meridian section of the spheroid, $\theta = 0$ at the top stagnation point, u_θ' is the tangential component of perturbation velocity,

$$X = \tfrac{2}{9}(2 - 3\cos\theta + \cos^3\theta),$$
$$Y = (\tfrac{1}{6}\pi)^{1/2}(\bar{\psi}/Ua^2)\chi^{1/2}R^{1/2},$$

and

$$S(X) = 3\chi^2/(1+\chi^2\cos^2\theta)^2.$$

The irrotational tangential velocity is $\bar{u}_\theta = 2U\chi\sin\theta/[\pi(1+\chi^2\cos^2\theta)^{1/2}]$. If χ is large, $S(X)$ can be approximated by a delta function in the form $\pi\chi\delta(X-\frac{4}{9})$, because $S(X)$ is of order χ^2 if $|\cos\theta| = O(1/\chi)$, a small region near $\theta = \frac{1}{2}\pi$ or $X = \frac{4}{9}$, and $S(X)$ is of order $1/\chi^2 \ll \chi^2$ if $|\cos\theta|$ is of order unity. Physically, this amounts to saying that the surface shear stress in the irrotational flow is concentrated at the equator, where $\bar{u}_\theta$ and its gradient are highest. It is a good approximation for positions downstream from there where $-\cos\theta \gg 1/\chi$. The rear stagnation vorticity is calculated as before, the equation analogous to (2.37) being

$$\begin{aligned} B(\bar{\psi}) &= b(m^2 s) \\ &= (3\sqrt{6}R^{1/2}U/4\pi\chi^{1/2}a^5)m^2 s\exp(-3Rm^4s^2/8\pi\chi^{5/3}a^6). \end{aligned} \tag{2.50}$$

We therefore find the ratio of perturbation to irrotational velocity at the surface as

$$\frac{u_\theta'}{\bar{u}_\theta} \doteqdot -\frac{\pi\chi^{11/6}\cos^2\theta}{R^{1/2}\sin^2\theta(-\cos\theta+\frac{1}{3}\cos^3\theta)^{1/2}}, \tag{2.51}$$

in the main part of the boundary layer on the rear half of the bubble, where $\pi-\theta \gg R^{-1/6}\chi^{-1/18}$, and

$$u_\theta'/\bar{u}_\theta \doteqdot -1.48\chi^{35/18}/R^{1/6}, \tag{2.52}$$

in the immediate neighborhood of the rear stagnation point where $\pi-\theta \ll R^{-1/6}\chi^{-1/18}$. Equation (2.52) gives $|u_\theta'/\bar{u}_\theta|>1$ for all values of R likely to occur in practice even at quite modest values of χ. With $\chi=3$, $R=600$, for example, we obtain $u_\theta'/\bar{u}_\theta=-4.3$ at the rear stagnation point, and $u_\theta'<-\bar{u}_\theta$ whenever $\theta>131°$, where $\sin\theta=0.76$. To put it another way, the perturbations are small only if

$$R \gg 10\chi^{35/3} > 10\chi^{11}. \tag{2.53}$$

It is tempting to identify the prediction of reversed flow at the rear of the bubble with separation of the boundary layer. But the numerical estimates above suffer from three important errors: we neglected higher terms in χ, higher boundary layer approximations, and deviations from spheroidal shape, and so the results can at best indicate general trends. Even if an exact theory predicted a region of reversed flow ($u_\theta<0$) on the surface, that alone would not show whether the back-eddy remained within the boundary layer or grew to a size comparable with the bubble. To answer that question would probably be as difficult and subtle as for flow past a rigid body, and no one has attempted it theoretically.

The question remains of why Moore's values of $C_D(R, M)$ are so insensitive to the breakdown in his model of the flow for large χ. If separation does not occur, but only a back-eddy inside a thin boundary layer, then his leading term (see Eq. 2.46) is unaffected although his second term involving $H(\chi)$ is wrong, and so one would not expect serious errors in C_D. If, on the other hand, separation does take place, then the back-eddy will contain slowly circulating fluid. In the irrotational flow for large χ the region behind the bubble where the back-eddy is situated would have contained small velocity gradients in any case; most of the viscous dissipation occurs near the equator where velocity gradients take their maximum values. In this case the error affects Moore's leading term, but not very much.

Recently, El Sawi (1970) has developed a new method of finding the function $W(\chi)$. He integrated the first moment of the inviscid equation of motion for a spheroid throughout the space occupied by fluid and showed that it yielded the required function. His maximum value of W was 3.271 instead of Moore's 3.745. El Sawi also considered bubble shapes slightly perturbed from either his or Moore's spheroids, and found the perturbed shapes closer to each other than the unperturbed, as one would hope. His best final result was that W tends to a limit near 3.2 as $\chi \to \infty$, instead of having a finite maximum at a finite χ as in the theories for spheroids. But an inviscid theory carried to this degree of precision is likely to be unrealistic in practice when viscous effects must be considered too.

E. Spherical-Cap Bubbles at High Reynolds Numbers

We have seen in the preceding section that two complications arise for large bubbles. Their boundary layers may separate and the hydrostatic term must be taken into account when evaluating the pressure in the fluid. As a result, a theoretical understanding of the motion is less complete than in the preceding cases. The first real advance was due to R. M. Davies and Taylor (1950), who related the speed of rise U to the radius of curvature r_c at the top stagnation point by the simple formula

$$U^2 = \tfrac{4}{9} g r_c . \tag{2.54}$$

This equation is semiempirical, being based on some experimental results which were not themselves explained theoretically. They are that the upper surface of the bubble is very nearly a cap of a sphere, and the pressure over that surface is very nearly the pressure calculated from Bernoulli's theorem for flow of an ideal fluid past the completed sphere of radius r_c, even though the wakes might be quite different. Then constancy of pressure inside the bubble requires constancy of

$$-\tfrac{1}{2}\rho(\tfrac{3}{2}U \sin \theta)^2 - g r_c \cos \theta + (2\sigma/r_c) \tag{2.55}$$

around its surface, where θ is the polar angle measured from the top stagnation point around the completed sphere. Equation (2.54) follows on expanding the expression (2.55) in powers of θ and equating the first non-trivial coefficient, of θ^2, to zero.

This simple theory agrees well with experiments, but it is not self-consistent (higher terms in θ do not vanish) and it does not predict other properties of the bubble and its wake. It is, nevertheless, better than the theory for large bubbles given by Levich (1962, Section 84). He ignored the hydrostatic pressure and made approximations which yielded a velocity independent of bubble size. That result occurs in practice only for bubbles in narrow tubes (Haberman and Morton, 1953, 1956; Peebles and Garber, 1953) but even then hydrostatic pressure determines the flow (R. M. Davies and Taylor, 1950).

1. *Theory for Low M*

Moore (1959) ignored surface tension for bubbles with large Reynolds and Weber numbers, and suggested a model for the flow analogous to Helmholtz's well-known free-streamline flow past a flat plate set across a stream (see Lamb, 1932). In Fig. 3, ABC is a cross section of the bubble,

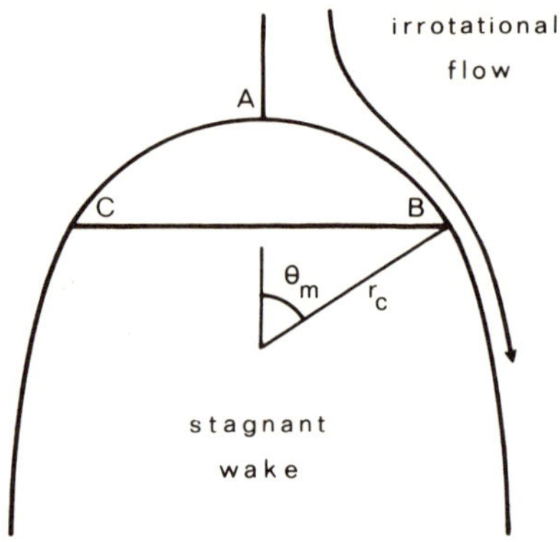

FIG. 3. The inviscid free-streamline model for spherical cap bubbles.

below which a wake containing fluid at rest extends to infinity. Outside the bubble and wake the flow is irrotational, with velocity U downwards at infinity. By Bernoulli's theorem, U is also the speed of the fluid at B and all points on the dividing streamline below B. Moore took the upper surface to be a spherical cap extending to $\theta = \theta_m$, and with Davies and Taylor's results (2.54) obtained the condition for equality of pressure at A and B in the form

$$gr_c(1 - \cos\theta_m) = \tfrac{1}{2}U^2 = \tfrac{2}{9}gr_c. \tag{2.56}$$

Hence $\cos\theta_m = 7/9$, or $\theta_m \doteqdot 39°$. The condition of R. M. Davies and Taylor (1950) that the flow be like that over the complete sphere then gives the velocity at B as $\tfrac{3}{2}U \sin\theta_m = 0.943U$, not the U required for consistency. This error of some 6% is a measure of the roughness of Moore's approximation to the shape and the flow field. If one proceeds with it, one finds the volume V to be

$$V = \tfrac{1}{3}\pi r_c^3(1 - \cos\theta_m)^2(2 + \cos\theta_m) = \tfrac{1}{6}\pi d^3, \tag{2.57}$$

and so the drag coefficient $C_D = 8gd/(3U^2) \doteqdot 1.95$.

Of course, a spherical cap is not the exact solution to the inviscid problem. We need instead that surface the irrotational flow past which satisfies $v^2 = 2gy$ on AB and has $v = U$ on the dividing streamline below B, where

v is the speed of the fluid and y the vertical distance below A. Rippin and Davidson (1967) calculated it numerically. They found that the shape is indeed close to a spherical cap, with $\theta_m = 50°$, that $C_D = 1.83$, and $U^2 = 0.62gr_c$. Moore's simple theory is thus a fair guide to the general form of Rippin and Davidson's more precise solution.

Unfortunately they both misrepresent the wake. Davies and Taylor's experiments were performed in nitrobenzene, which has an optical anisotropy allowing one to see where the fluid is strongly sheared. They found a turbulent region under the bubble, filling the remainder of the sphere defined by its upper surface. Maxworthy (1967) used more direct flow-visualization techniques and showed that there is also a turbulent wake, of the same order of width as the bubble, extending a considerable distance below it. The turbulence is not surprising: steady flow past bubbles of the size used by Davies and Taylor or by Maxworthy would have been highly unstable.

Before Maxworthy had shown that the wake was not closed if turbulent, Collins (1966) had tried a model for the flow in which the bubble and its recirculating wake formed a closed surface whose spherical polar equation was assumed to be $r = a_0(1 - \varepsilon \sin^4 \theta)$. He obtained $\varepsilon = 0.0785$ by requiring the flow past the surface to be irrotational and making it satisfy the pressure condition near $\theta = 0$ to order θ^4 instead of Davies and Taylor's θ^2 only. The upper part ($\theta < 36°$) of Collins's assumed surface closely resembles a spherical cap with radius $\bar{a} = 0.953a$. Collins's result is $U^2 = 0.425g\bar{a}$ instead of Davies and Taylor's $0.444g\bar{a}$, and it fits the experiments somewhat better. It is interesting to test the way in which the results depend on the function of θ assumed for r. If one repeats Collins's calculation with the well-known irrotational flow past a spheroid (see Lamb, 1932), one finds that the spheroid must be oblate, with eccentricity $\frac{1}{2}$, so that the ratio of equatorial to polar diameter is $2/\sqrt{3} = 1.155$ (cf. Collins's 0.9215), and $U^2 = 0.414g\bar{a}$ (which fits the experiments about as well as Collins's 0.425).

One can therefore obtain fairly good estimates of $U^2/g\bar{a}$ from theories of this sort, but they do not readily yield information about the true shape of the wake, not even whether it is elongated or flattened from a sphere. In addition, closed-wake models which ignore viscosity and surface tension cannot possibly give the size of the bubble correctly. Figure 4 shows why. If ABCD represents a cross section of the bubble and BCDE its wake, the inviscid flow must be more or less as shown. A, C, and E are all stagnation points and so Bernoulli's theorem gives the pressure difference $p(\mathrm{C}) - p(\mathrm{A})$ as $\rho g \times \mathrm{CA}$. But if surface tension and the density of the gas can be neglected, $p(\mathrm{C}) = p(\mathrm{A})$ and so the bubble must have zero height.

This objection does not apply to Moore's model as improved by Rippin and Davidson. which is self-consistent for zero surface tension and viscosity. But to make the solution unique one must impose the somewhat artificial condition that U have the largest possible value for a given bubble

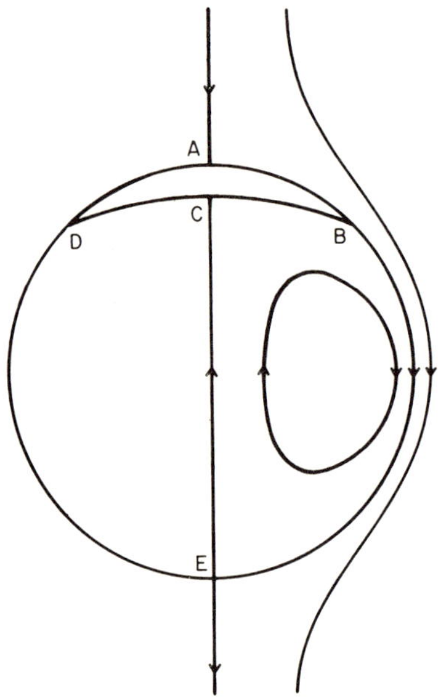

FIG. 4. The closed-wake model for spherical cap bubbles.

volume, as was shown by Garabedian (1957). The other artificial feature of Rippin and Davidson's model is its stagnant wake with a vortex sheet dividing it from the flow outside. Such sheets are well known to be highly unstable, and the wake is really turbulent (Maxworthy, 1967). A comparison of stagnant-wake theories and experiments for a variety of bodies reveals, however, that they give useful approximations to the flow field over the front part of the bodies, and underestimate the drag consistently. For axisymmetric bodies (sphere, disk, bubble) the stagnant-wake theory gives some 70% of the true drag, and for two-dimensional ones (flat plate at 90° incidence, circular cylinder) about 50%, as shown in Table 2. This table was compiled from data given by Birkhoff and Zarantonello (1957), Rippin and Davidson (1967), R. M. Davies and Taylor (1950), Perry *et al.* (1963), and Prandtl and Tietjens (1957).

Recently, Davenport *et al.* (1967a) have confirmed Davies and Taylor's value of C_D (2.65) in liquids of lower M than in any previous experiments: mercury ($M = 7.3 \times 10^{-14}$) and molten silver (3.5×10^{-13}). Shapes of bubbles were determined only in mercury, where bubbles of a given volume had the same height as in water but smaller equatorial radii. As Davenport *et al.* concluded, surface tension must have rounded off the sharp "corner"

TABLE 2

DRAG COEFFICIENTS FOR VARIOUS BODIES[a]

Body	$C_D = C_{DS}$ (stagnant wake)	$C_D = C_{DE}$ (experiment)	C_{DS}/C_{DE}
Spherical cap bubble	1.83	2.65	0.69
Sphere	0.3	0.45	0.7
Disk across stream	0.8	1.1	0.7
Flat plate across stream (two-dimensional flow)	0.88	2.0	0.44
Circular cylinder (two-dimensional flow)	0.61	1.1	0.55

[a] Experiments for subcritical flow, i.e. laminar boundary layer upstream of separation, but turbulent wake. The numerical values are from R. M. Davies and Taylor (1950), Prandtl and Tietjens (1957), Birkhoff and Zarantonello (1957), Perry *et al.* (1963), and Rippin and Davidson (1967).

at B in Fig. 3 or 4 to a greater radius. The bubbles were therefore really in the transitional range from spheroids to spherical caps, and if the mercury had been pure one would have expected lower values of C_D. Haberman and Morton's (1953) figure 29 shows that drag coefficients closer to those of spherical caps are to be expected in impure liquids.

2. *Theory for High M*

For bubbles in liquids with high values of M one cannot appeal to turbulence in the wake to account for the drag. Most experimenters have found the motion to be steady and stable (see Hartunian and Sears, 1957), though not Davenport *et al.* (1967b). If Hartunian and Sears are right in suggesting that the motion in a pure liquid is always stable if $W > 3.18$ for the bubble with $R = 200$, we can deduce a critical value of M from Moore's (1965) theory of the drag. The criterion is that any bubble should be stable in any liquid with $M > 1.4 \times 10^{-8}$. It is a pity that so few experiments have been done with M of order 10^{-7} to test this theory. Davenport *et al.* (1967b) found slight instability with M values up to 1.6×10^{-2} in polyvinyl alcohol solutions. One wonders whether some property of this surface-active macromolecular solute could have caused it.

Wakes in high-M liquids are quite unlike the turbulent chaotic eddies seen by Maxworthy (1967). Slaughter and Wraith (1968) gave a good photograph of a spherical-cap bubble rising in a glycerol solution with M probably between 10^{-2} and 10^{-3} (not stated by the authors). The wake was a toroidal vortex resembling Hill's (1894) spherical vortex, but somewhat elongated in the direction of flow, and followed by a wake of the usual type in laminar flow far behind an axially symmetric body (Rosenhead, 1963) in which the velocity perturbation gradually decays to zero but the momentum defect remains finite. The only theory so far available for such flows is due to Parlange (1969), who exploited the analogy between the circulating wake and the circulating interior of a drop of one fluid moving in another (see Section III). He assumed that the bubble has little effect on the dynamics of the flow in the wake (except causing it to exist!), so that the vortex was near enough to a sphere for Harper and Moore's (1968) drag theory to apply. If so,

$$\tfrac{4}{3}\pi\rho a^3 g = 30\pi\eta U r[1-(6.6-0.14\ln R'/R'^{1/2})], \tag{2.58}$$

where a is the equivalent spherical radius as before, r the radius of the wake sphere, and $R'=2Ur/\nu$. Parlange also assumed R. M. Davies and Taylor's (1950) relation, $U^2=\tfrac{4}{9}gr$ (see Eq. 2.54). From these equations and the definition of C_D (Eq. 2.3) one deduces that

$$R=\tfrac{1}{6}R'C_D, \tag{2.59}$$

and

$$C_D=6\left[\frac{20}{R'}\left(1-\frac{6.6-0.14\ln R'}{R'^{1/2}}\right)\right]^{1/3}, \tag{2.60}$$

and finds C_D and R in terms of R'. The resulting graph of $C_D(R)$ is shown in Fig. 1 as curve I, down to $R'=100$, below which value Harper and Moore's theory must be seriously in error. It is evidently in the right part of the (R, C_D) plane at this point ($R=44.6$, $C_D=2.56$), and the trend for C_D to decrease slowly as R increases is experimentally plausible. No experimental test of the asymptotic form for large R has yet been published; the prediction is that $C_D \sim 26.8R^{-1/2}$ as $R\to\infty$. Angelino's (1966) results do not inspire much confidence in it, but they cease at $R=300$.

Parlange's theory is at best a first approximation for high R, when the bubble occupies a very small fraction of the wake sphere. It is debatable whether the logarithmic term is worth including in Eqs. (2.58) and (2.60), because Harper and Moore's arguments for including it in the theory of drops (see Section III, B, 2) do not apply when a bubble occupies the top stagnation region.

Parlange's assumption that the closed-wake region should be nearly spherical seems easier to justify; Hill's spherical vortex is the only known simply connected shape for which the irrotational flow outside and the flow with constant circulation density Ω inside can have the same surface speeds. Ω must be constant inside the "wake sphere," by the Prandtl-Batchelor circulation theorem (Prandtl, 1905; Batchelor, 1956). It seems reasonable to suppose that in Parlange's case, where the bubble occupies only a small part of the "wake sphere," the shape of the latter would be slightly perturbed from Hill's vortex. To find these perturbations will not be easy. The shape will depend on the differences between internal and external velocities which give rise to dynamic pressure differences of order $\rho U^2 R^{-1/2}$ (Harper and Moore, 1968). In many experiments these are of the same order as surface-tension pressures, σ/r, and neither may be neglected. Furthermore, much of the bubble lies within the stagnation regions, whose theory we have already seen to be very complicated in Section II,C.

3. *Skirts*

In some liquids with very high values of M (0.2 or more), spherical-cap bubbles develop thin "skirts" trailing downwards (Jones, 1965; Davenport *et al.*, 1967b; Guthrie and Bradshaw, 1969), as shown schematically in Fig. 5. The thickness of the skirt was calculated by Guthrie and Bradshaw (1969) by equating the pressure rises down the skirt in the gas inside it

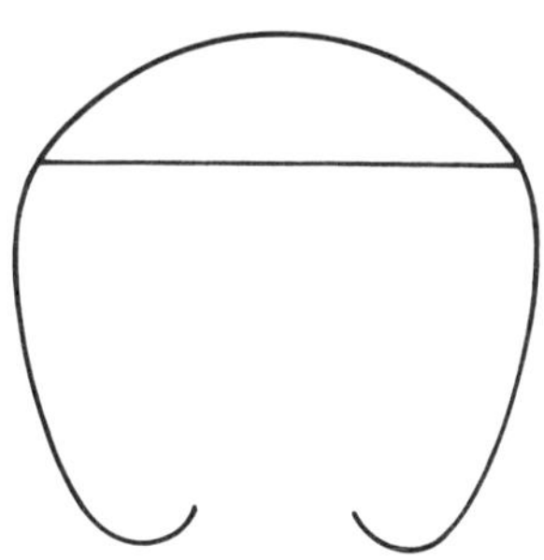

FIG. 5. Cross section of a spherical cap bubble with trailing skirt, after Guthrie and Bradshaw (1969).

and the liquid outside. The gas was treated as a two-dimensional Poiseuille flow and the hydrostatic pressure was assumed to dominate in the liquid. Guthrie and Bradshaw's experimental bubbles of volume 53 cm^3 had skirts 41 μm thick. Their theory gave thicknesses of 54 μm for stagnant liquid inside the skirt or 76 μm for the same liquid velocity inside and out.

It is not clear what determines whether a skirt is present. Davenport *et al.* (1967b) found them on sufficiently large bubbles in polyvinyl alcohol solutions with $M = 0.185$, but Jones (1965), who observed them in glycerol with $M = 6720$ did not observe them in mineral oil with $M = 318$. Nor are the lengths of the skirts well understood. Davenport *et al.* observed a steady length and shape resembling Fig. 5, Jones found them to be shorter and variable, and Astarita (1970) has had them extend right down the path of the bubble.

F. The Wave Analogy

Mendelson (1967), Cole (1967), and Malenkov (1968) have recently tried to draw an analogy between spherical-cap bubbles rising in low-M liquids and waves traveling along a plane free surface. The motion of both waves and bubbles depends mainly on g, ρ, and σ, viscosity being unimportant to first order (but, see, Walbridge and Woodward, 1970), and in both cases the problem is to predict a velocity as a function of a length scale (bubble size in one case, wavelength in the other). But the theory of surface waves is very much easier.

If one identifies the speed of rise of the bubble with the phase velocity of the wave given by Kelvin (see Lamb, 1932) as

$$U^2 = \frac{g\lambda}{2\pi} + \frac{2\pi\sigma}{\lambda\rho}, \tag{2.61}$$

where λ is the wavelength, it remains to ascertain which parameter of the bubble is best to call λ. R. M. Davies and Taylor (1950) give $U^2 = \frac{1}{2}gd$ as a good approximation for large bubbles, and so Mendelson (1967) suggested that λ should be put equal to the equatorial circumference of the equivalent sphere, πd. This substitution yields graphs of U against d which agree reasonably well with the rather scattered data, but it does not give a smooth transition between the limit of large spheroidal bubbles (W constant near 3) and spherical caps (C_D constant near 8/3). This is because Eq. (2.61) with Mendelson's substitution $\lambda = \pi d$ gives

$$C_D = \frac{8}{3}\left(1 - \frac{1}{2W}\right), \tag{2.62}$$

after a little algebra using Eqs. (2.3) and (2.6). Clearly (2.62) is inconsistent with C_D varying while W has any constant value. The wave theory therefore does not offer a short cut to the very difficult theory of the spheroid–spherical cap transition.

III. A Drop with Constant Surface Tension Moving Under Gravity

A. Introduction

The theory of moving liquid drops is in most respects very like that of gas bubbles, but more complicated. That is why bubbles were described separately in Section II. Instead of one (M), there are now three independent dimensionless parameters which characterize the liquids, inside and outside the drop. These are ρ_0/ρ_1, η_0/η_1, and

$$M_0 = g\eta_0{}^4|\rho_0 - \rho_1|/\rho_0{}^2\sigma^3, \tag{3.1}$$

the reciprocal of Hu and Kintner's (1955) parameter P. Here σ is the interfacial tension and g the acceleration due to gravity as before, and ρ_i and η_i are the density and viscosity, for $i = 0$ outside the drop (the continuous phase) and for $i = 1$ inside it (the dispersed phase). Guided by Section II,A we define the Reynolds number R_0, drag coefficient C_{D0}, and Weber number W_0 by the equations

$$R_0 = Ud/\nu_0 = Ud\rho_0/\eta_0, \tag{3.2}$$

$$C_{D0} = \text{force on drop}/\tfrac{1}{2}\rho_0 U^2\pi a^2 = 4gd|\rho_0 - \rho_1|/3\rho_0 U^2, \tag{3.3}$$

$$W_0 = \rho_0 U^2 d/\sigma = (4M_0 R_0{}^4/3C_{D0})^{1/3}. \tag{3.4}$$

As before, $d = 2a$ is the diameter of the sphere with the same volume as the drop (the equivalent sphere). The modulus of $\rho_0 - \rho_1$ appears in the formulas because it is convenient to have positive drag coefficients whether the drop moves up or down. One can define M_1, R_1, C_{D1} and W_1 by interchanging subscripts 0 and 1 in Eqs. (3.1)–(3.4), but the parameters M_0, R_0, C_{D0}, and W_0 are more often useful.

Figure 1 illustrates the behavior of bubbles, but would serve also for drops with very large values of both ρ_0/ρ_1 and η_0/η_1. We shall not attempt to draw the corresponding graphs for general values of ρ_0/ρ_1 and η_0/η_1, but instead indicate how the theories of bubble motion must be modified for drops. As we shall see, information is even less complete than for bubbles.

B. Theory for a Spherical Drop

1. *Low Reynolds Number*

The analysis of Section II,C,1 for bubbles follows the original work first of Hadamard (1911) and Rybczynski (1911) and then of Taylor and Acrivos (1964). It presents the special case $\eta_1 = \rho_1 = 0$ of the same authors' theories of drops, which use the same methods. The boundary conditions at the surface are now continuity of tangential velocity and shear stress, and zero values of normal velocity both inside and out. Uniqueness of the solution for sufficiently small R_0 is ensured by the additional conditions that the velocity be a finite continuous function of position at the center $r = 0$, and that it tend to a constant vertical vector of magnitude U as $r \to \infty$, downwards if $\rho_1 < \rho_0$, upwards if $\rho_1 > \rho_0$.

Equations (2.7)–(2.9) still describe the flow both inside and out, in the limits $R_i \to 0$, $rR_0 \to 0$, and Eq. (2.10) is still the general solution outside the drop. Inside it, the stream function ψ is given by

$$\frac{\psi}{Ua^2} = \sum_{n=2}^{\infty} \left(\gamma_n \frac{r^{n+2}}{a^{n+2}} - \delta_n \frac{r^n}{a^n} \right) \mathscr{I}_n(\mu), \tag{3.5}$$

if it is (without loss of generality) required to vanish at the origin. As before, $\mathscr{I}_n(\mu)$ is the integrated Legendre polynomial of Eq. (2.11). The coefficients γ_n, δ_n must be evaluated from the boundary conditions. For a spherical bubble $\gamma_n = \delta_n$ for all n, and continuity of tangential velocity and shear stress at $r = a$ gives, on substitution from (2.10) and (3.5),

$$\begin{aligned} \frac{u_\theta \sin\theta}{U} &= \frac{1}{Ua}\frac{\partial \psi}{\partial r} \\ &= (3 - 2\alpha_2)\mathscr{I}_2(\mu) - \sum_{n=3}^{\infty} 2\alpha_n \mathscr{I}_n(\mu) \\ &= 2\sum_{n=3}^{\infty} \gamma_n \mathscr{I}_n(\mu), \end{aligned} \tag{3.6}$$

$$\begin{aligned} \frac{p_{r\theta}\, a \sin\theta}{U} &= \frac{\eta_i}{U}\frac{\partial}{\partial r}\left(\frac{1}{r^2}\frac{\partial \psi}{\partial r}\right) \\ &= \eta_0\left\{6(\alpha_2 - 1)\mathscr{I}_2(\mu) + \sum_{n=3}^{\infty} (4n-2)\alpha_n \mathscr{I}_n(\mu)\right\} \\ &= \eta_1 \sum_{n=2}^{\infty} (4n-2)\gamma_n \mathscr{I}_n(\mu), \end{aligned} \tag{3.7}$$

for all $\mu = \cos\theta$ in $-1 \leq \mu \leq 1$. We may now equate coefficients of $\mathscr{I}_n(\mu)$, and find $\alpha_n = \gamma_n = 0$ for $n \geq 3$ as before, while $\alpha_2 = (2\eta_0 + 3\eta_1)/(2\eta_0 + 2\eta_1)$, $\gamma_2 = \eta_0/(2\eta_0 + 2\eta_1)$. Hence

$$C_{D0} = 8(2\eta_0 + 3\eta_1)/R_0(\eta_0 + \eta_1), \tag{3.8}$$

and

$$u_\theta = U\eta_0 \sin\theta/2(\eta_0 + \eta_1) = v_0 \sin\theta, \tag{3.9}$$

on $r = a$. As $\eta_0/\eta_1 \to \infty$ we recover the previous results for bubbles: $C_{D0} \to 16/R_0$ and $v_0 \to \frac{1}{2}U$. As $\eta_0/n_1 \to 0$ we recover Stokes's results for rigid spheres: $C_{D0} \to 24/R_0$ and $v_0 \to 0$. For intermediate values of η_0/η_1, C_{D0} and v_0 take intermediate values.

Inside the drop the stream function and hence the velocity components are simple fixed functions of position, multiplied by v_0. One streamline pattern will therefore serve for any value of v_0, and it is shown in Fig. 6. It is given by

$$\psi = \tfrac{1}{4}v_0[(r^4/a^2) - r^2]\sin^2\theta \leq 0 \qquad (3.10)$$

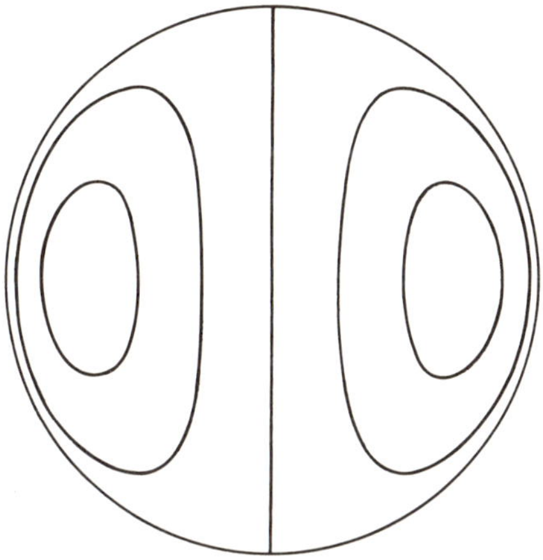

FIG. 6. Streamline pattern inside Hill's spherical vortex.

inside the drop. The flow goes round the surface and back up the middle. The maximum speed is v_0 at the equator and the center, and there is a stagnation ring at $r = a/\sqrt{2}$, $\theta = \frac{1}{2}\pi$. This flow pattern was discovered by Hill (1894) as a solution of the equations of inviscid motion; it satisfies the Navier–Stokes equations exactly for any value of the Reynolds number if the boundary conditions are suitable. So does Hill's (1894) spheroidal vortex whose stream function is

$$\psi = Ar^2 \sin^2\theta(Br^2 \cos^2\theta + Cr^2 \sin^2\theta - 1),$$

where A, B, C are constants. Both vortices appear below in the theory for high Reynolds numbers. The internal motion has been elegantly observed by Tyroler *et al.* (1971) and its general features confirmed, although their Reynolds numbers were not small enough to make the theory strictly applicable.

The theory above holds only in the Stokes flow limit $R_0 \to 0$. Higher terms in the expansion for small R_0 were given by Taylor and Acrivos (1964). We quote here their result for $R_0 C_{D0}$, but only to order R_0:

$$R_0 C_{D0} = \frac{8(2\eta_0 + 3\eta_1)}{\eta_0 + \eta_1}\left(1 + \frac{R_0(2\eta_0 + 3\eta_1)}{16(\eta_0 + \eta_1)} + O(R_0^2 \ln R_0) + O(W_0)\right). \tag{3.11}$$

Taylor and Acrivos also gave the coefficients of $R_0^2 \ln R_0$ and W_0, but not that of R_0^2 which often contributes to $R_0 C_{D0}$ to the same order. The practical value of these higher terms is in any case disputable for solid spheres (see Chester and Breach, 1969; Pruppacher *et al.*, 1970) and we have no reason to believe the situation is better for fluid spheres. To take many terms of a divergent asymptotic series only gives a close approximation to its sum very near the point where a small number of terms would be adequate.

The shape of the drop was also investigated by Taylor and Acrivos. To leading order, the axis ratio χ is given by

$$\chi - 1 = \frac{3W_0}{16(\kappa+1)^3}\left[\left(\frac{81}{80}\kappa^3 + \frac{57}{20}\kappa^2 + \frac{103}{40}\kappa + \frac{3}{4}\right) - \frac{\rho_1 - \rho_0}{12\rho_0}(\kappa+1)\right] + O(W_0 R_0) + O(W_0^2 R_0^{-1}), \tag{3.12}$$

where $\kappa = \eta_1/\eta_0$. The drop is, to first order, a spheroid, but it may be either prolate or oblate, unlike a bubble. Prolate (elongated) shapes require the inner fluid to be the denser, and either less viscous than the outer fluid, or else very little more. Liquid drops in gases, for example, are normally oblate, as are mercury drops in water. In the limit $\kappa \to \infty$ (drop behaving like a solid sphere), $R_0 C_{D0} \to 24(1 + \frac{3}{8}R_0)$ and $\chi - 1 \to (243/1280)W_0 \doteqdot 0.190W_0$. As $\chi - 1 \doteqdot 0.156W_0$ for a gas bubble, the shape is not very sensitive to the interior viscosity if the drop rises and so $\rho_1 < \rho_0$.

2. *High Reynolds Number*

If the Reynolds number is large, the analogy between drop and bubble motion is less close than for small R_0. Internal circulation cannot be neglected for drops, and it brings fluid particles repeatedly around closed loops in steady motion. We must therefore impose the condition that the circulation density Ω of each fluid particle returns to its original value on completing the path. Inside the drop and away from the thin boundary layers, that requires Ω to be constant if R_0 is large, by Prandtl's (1905) and Batchelor's (1956) theorems.

But if Ω is constant inside a sphere the flow must be Hill's spherical vortex (Eq. 3.10), with surface velocity, $v_1 \sin\theta = (2\Omega a^2 \sin\theta)/5$. We therefore envisage the motion of the drop as follows, as it starts from rest. Initially there is no motion inside the drop ($\psi = \Omega = 0$), and irrotational flow with surface velocity $v_0 \sin \theta = \frac{3}{2}U \sin \theta$ outside. A boundary layer must develop around the surface because of this velocity discontinuity, and the circulation density Ω within it will be carried along with the fluid and diffused across streamlines. The beginning of this process is described by Sumner and Moore (1969), who gave the first two terms of an expansion in powers of the time. But the boundary layer cannot remain a transition region between irrotational flow and none. The circulation density in the layer must be of order $U/(\delta a)$ initially both inside and outside the surface if η_0 and η_1 are of the same order and the surface tension is constant, and δ is the boundary layer thickness. Ω will then diffuse both inwards and outwards until in a steady state reached after a time $O(a^2/\nu)$ it has the same order of magnitude throughout the interior, the boundary layer, and the wake. In the interior it is then of order v_1/a^2 and in the boundary layer $(v_0 - v_1)/(\delta a)$, and so

$$v_0 - v_1 = O(v_1 \delta a/a^2) = O(v_1 \delta/a) \ll v_1, \tag{3.13}$$

if the Reynolds number is high.

To a first, inviscid approximation we may take $v_0 = v_1$ if R_0 is high, the flow steady, the drop spherical, the surface tension constant, and the interior viscosity of the same order as the exterior. Sumner and Moore (1969) argue against this conclusion, but prove only that denying any of the assumptions invalidates it. A separate question is whether the assumptions can all be true for real drops in real fluids. We investigate this more fully below; the answer is that they can, to a good approximation, but only for a restricted class of drops.

The importance of spherical shape to the argument is interesting. It enables the tangential velocity to be continuous across the surface (as Hill realized in 1894), unlike any other known simply-connected closed surface. No proof has ever been given that other such surfaces do not exist, however. Recently Fraenkel (1970) suggested that doubly-connected ones do. He found a one-parameter family of toroidal vortices with Ω constant inside, irrotational flow outside, and velocity continuous everywhere.

Having arrived at a velocity distribution which is continuous across the surface of the drop for the limiting case $\sigma \to \infty$, $R_0 \to \infty$ (in that order), we can now use the methods described for spherical bubbles in Section II,C,2 to analyze the boundary layers. As in the case of bubbles, these layers must exist because the inviscid velocity field does not satisfy the tangential stress condition, and the velocity perturbation is of order $\delta/a = O(R^{-1/2})$ of the inviscid velocity. But there is a major complication. The boundary layers merge into the rear stagnation regions inside and outside the drop, and the fluid in them turns through 90° to travel up and down the vertical axis of symmetry. Outside the drop, this fluid forms the upstream end of the wake, but inside it, it travels with very little diffusion of circulation density (see Harper and Moore, 1968) until it rejoins the boundary layer at the front stagnation point (see Fig. 7). The distribution of Ω therefore cannot

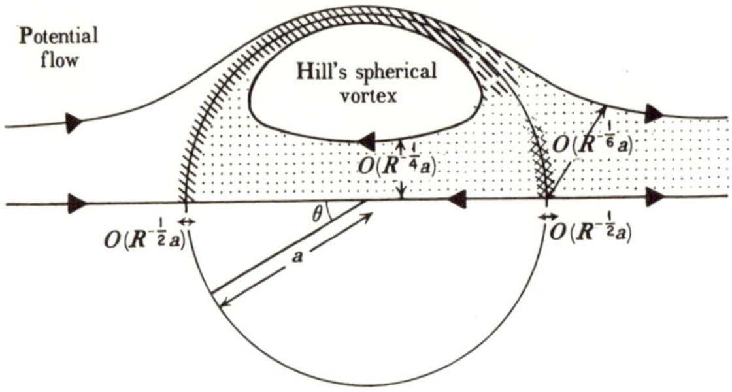

FIG. 7. The form of the boundary layers and wakes of a spherical drop at high Reynolds numbers. Diagonally shaded: the stress-induced viscous boundary layer. Dotted: the essentially inviscid stagnation regions and wakes. Cross-hatched: the inner viscous boundary layer near the rear stagnation point. [From Harper and Moore (1968), by permission of the Cambridge University Press.]

be specified at the front stagnation point without solving the problem of its diffusion in the boundary layer. One has to let Ω be some arbitrary function of ψ inside the drop at the front stagnation point, follow its variation around the surface and back up the middle, and use the condition that the starting and finishing values be the same.

Similar considerations apply (Brignell, 1970) to convective diffusion of a solute. Boundary layer analyses which ignore the coupling between rear and front stagnation regions of a drop must be in error, whether they purport to describe the distribution of circulation density or of solute, unless the effects of internal diffusion are negligible or the motion has been going for such a short time that fluid from the rear stagnation region has not yet reached the front. This fact was realized, though not corrected for in the theory, by Ruckenstein (1967) and Taunton and Lightfoot (1969). It was unfortunately ignored by Winnikow and Chao (1966).

The closed-loop condition on Ω was studied by Harper and Moore (1968). It can be written as an integral equation to be solved for a function $g(z)$ in $0 \leq z < \infty$, namely

$$\pi^{-1/2} \int_0^\infty g(z')\{\exp[-(z-z')^2] + \lambda_a \exp[-(z+z')^2]\}\, dz' = g(z) + h(z), \tag{3.14}$$

where λ_a is a parameter between -1 and 1, and $h(z)$ is a known function. Special cases of this equation, which also appears in magnetohydrodynamics and the theory of evolution of comet orbits, have been studied by Hammersley (1961), Kendall (1961), and Stewartson (1968) for $\lambda_a = 0$, and by Kochina (1969) for $\lambda_a = -1$.

In Harper and Moore's paper, z is defined to be $3R_1^{1/2}(a - r)\sin^2 \theta/8a$. This differs from Eq. (2.31) above by the substitution of R_1 for R_0 and by its sign. Both changes make it more convenient for use inside the drop. The values of λ_a, $g(z)$, and $h(z)$ are

$$\lambda_a = (1 - V')/(1 + V'), \tag{3.15}$$

$$g(z) = \frac{4\sqrt{2}a^2}{3U} \int_z^\infty \Omega'(z')\,dz', \tag{3.16}$$

$$\begin{aligned} h(z) &= \frac{2\sqrt{2}(2V' + 3)}{V' + 1} \lambda_b \operatorname{ierfc}(z) \\ &\quad + \frac{3V'c}{V' + 1} \int_{-1}^{1} \mu \operatorname{erfc}\left(\frac{2z}{(1 + \mu)(2 - \mu)^{1/2}}\right) d\mu \\ &= \lambda_b g_b(z, \lambda_a) + C g_c(z, \lambda_a), \end{aligned} \tag{3.17}$$

say, where $\Omega'(z)$ is the perturbation circulation density in a stagnation region, $V = \eta_0/\eta_1$, $V' = (\eta_0 \rho_0/\eta_1 \rho_1)^{1/2}$, and $\lambda_b = (2V + 3)/(2V' + 3)$. The expression for $g_c(z, \lambda_a)$ is simpler than that of Harper and Moore but is equivalent to it. C in Eq. (3.17) is related to the strength of the Hill's vortex, which is $1 + (2/R_1)^{1/2}C$ times its unperturbed value. One finds C from the condition that the tangential velocity perturbation in a boundary layer must tend to zero as the layer merges into effectively inviscid flow at its outer limit, i.e. $g(z) \to 0$ as $z \to \infty$. Only one value of C will give that result in Eqs. (3.14) and (3.17), and it is negative. This corresponds to the internal vortex being slowed down by viscosity, which seems reasonable. The values of C/λ_b are within 2% of $-2.5(2 + V')/V'\sqrt{2}$. With that approximation the internal circulation is less than that of Hill's classical vortex by a factor of

$$1 + \frac{C\sqrt{2}}{R_1^{1/2}} = 1 - \frac{2.5(2V + 3)(V' + 2)}{R_0^{1/2} V(2V' + 3)} \tag{3.18}$$

More precise values of C/λ_b will be found in Table 3, with data on the drag.

With that major complication of allowing for recirculation out of the way, there is now a minor one to be considered. The theory of the stagnation regions predicts velocity discontinuities of order $UR_i^{-1/3}$ developing at distances of order $aR_i^{-1/6}$ from the stagnation points. Such discontinuities in an inviscid model imply viscous boundary layers in the real flow. Near the rear stagnation point this layer develops after the main viscous layer has died out, but it turns out to be of little dynamical significance. Near the

TABLE 3

DRAG FUNCTIONS FOR SPHERICAL BUBBLES[a]

V'	0.2	0.5	1.0	2.0	5.0	∞
$3\lambda_a$	2	1	0	−1	−2	−3
c_1	0.0275	0.0935	0.177	0.262	0.335	0.390
c_2	8.89	7.56	6.15	5.15	4.22	3.56
c_3	−9.72	−9.00	−8.22	−7.41	−6.59	−5.77
C/λ_b	−19.86	−8.900	−5.282	−3.500	−2.452	−1.772

[a] The second-order terms c_1, c_2, c_3 in Eq. (3.18) for the drag coefficient, and the internal flow perturbation parameter C/λ_b, as functions of λ_a or of V', copied from Harper and Moore (1968) by permission of the Cambridge University Press.

front it merges into the main layer. It does not appreciably affect Eq. (3.14) but it does contain viscous dissipation of energy at a rate sufficient to add a term $O(R_0^{-3/2} \ln R_0)$ to the drag coefficient. The other terms in the drag are calculated from the viscous dissipation in the irrotational flow, the internal vortex, the boundary layers, and the wake as for bubbles, and the final result is

$$C_{D0} = \frac{48}{R_0}\left\{1 + \frac{3}{2V} + \frac{\lambda_b(V'+1)}{R_0^{1/2}V}\left[\frac{\lambda_b c_1(\lambda_a)\ln R_0}{V} + \frac{\lambda_b(1+V')c_2(\lambda_a)}{V} + \left(1+\frac{3}{2V}\right)c_3(\lambda_a)\right]\right\}, \tag{3.19}$$

where c_1, c_2, and c_3 are functions of λ_a given in Table 3. The approximations made near stagnation points have the effect on C_{D0} of producing errors of the same order as the logarithmic term (which is, fortunately, small). It is therefore worthwhile when comparing the theory with experiments, or predicting drop velocities, to do the calculation both with and without that term.

Such an elaborate theory would not be worth using if it did not agree better with experiments than the available alternatives. It does in carefully purified liquids, provided that M_0 is low enough for drops to remain nearly spherical up to Reynolds numbers of several hundred and provided that the calculated perturbations are reasonably small. The values of C_{D0} then come within 20% of the experimental data, whereas C_{D0} for a rigid sphere is a factor of 2 too high and Winnikow and Chao's (1966) over-simple theory is a factor of 2 or 3 too low. The experiments in question were done by Licht and Narasimhamurty (1955), Elzinga and Banchero (1961), Winnikow and Chao (1966), and Thorsen *et al.* (1968). Graphs of C_{D0} against R plotted logarithmically on both axes are presented in Fig. 8. The curves are generally similar to those for bubbles in low-M_0 liquids, except for two liquids in which there was a sharp break in the curves at about $R_0 = 400$, due probably to surface contamination.

Agreement between experiment and theory is worse than for bubbles in general, although it is better than for bubbles in water. All the drop experiments had water as continuous phase, except one which had it as dispersed phase. Perturbations predicted in the theory are often larger fractions of the first-order terms than for bubbles, especially if the dispersed phase is the more viscous; if $\eta_1 > 2\eta_0$ the theory is unlikely to be applicable. But Sumner and Moore (1969) exaggerate when they suggest that it is useless whenever η_1 is of the same order as η_0.

The theory is also inapplicable if the first-order drag coefficient $48(1 + 3/2V)/R_0$ is larger than that of a rigid sphere at the same R_0, or if the predicted internal circulation is less than about half of its value in the inviscid theory (see Eq. 3.18). One then gets a better result by assuming the drop to behave as if rigid.

C. Distortions from Sphericity

If R_0 is high enough, drops, like bubbles, will not be spherical. Experimentally, the effects on drops and bubbles are rather similar; see Winnikow and Chao (1966) and Thorsen *et al.* (1968). As R_0 increases C_{D0} passes a minimum value and then increases; at slightly higher R_0 the drops begin to oscillate (either sideways like bubbles or oblate-to-prolate while rising vertically). Then, if it was initially well below the solid sphere value, C_{D0} becomes proportional to R_0^4, as W_0 is constant at about 4 (making $C_{D0} \doteqdot M_0 R_0^4/48$) until about $C_{D0} = 1$. Beyond that value C_{D0} rises rather less steeply, but the curves do not level off to a constant value as they do for bubbles. Oscillating falling drops emitted from nozzles are so much less stable than rising bubbles released from dumping cups that they break up when W_0 is of order 10 (see Lane and Green, 1956). It is a pity that no data seem to be available for rising drops in low-M_0 fluids at high Weber numbers nor for drops released from dumping cups. (Bubbles from nozzles may also break up.)

One might hope that a theory like that of Moore (1965) could account for the deformations of drops as successfully as it has for bubbles. Unfortunately, the flow past a spheroid can never be slightly perturbed from a condition of irrotational outside and constant Ω inside unless the shape is

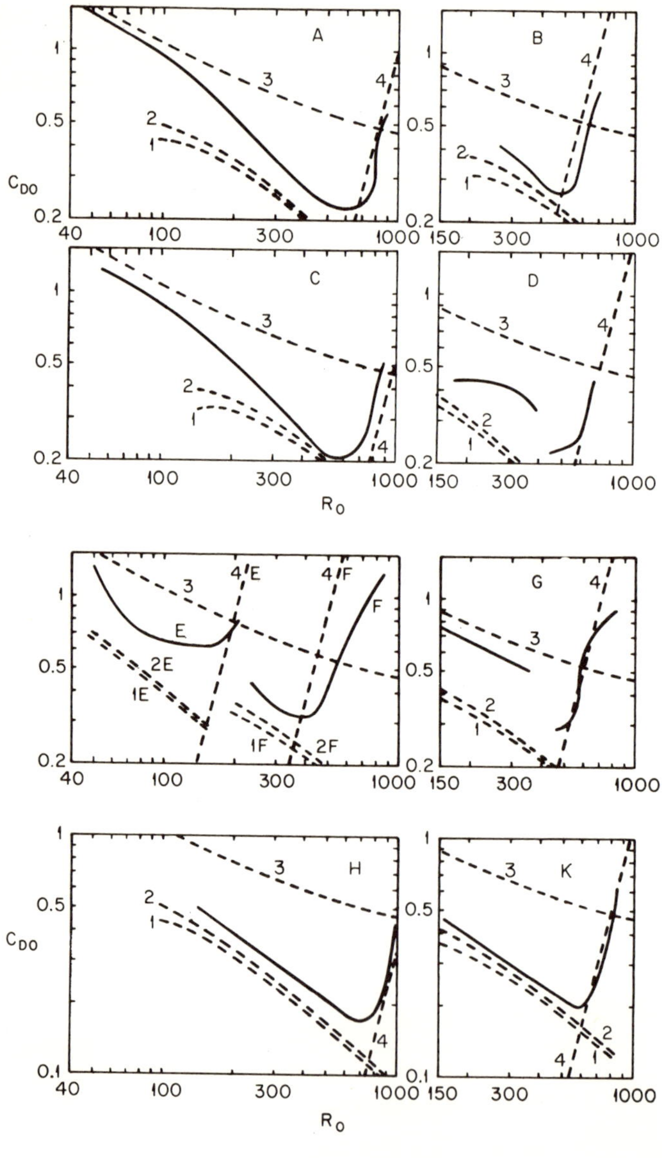

very close to a sphere. The reason, given by Hill (1894), is that the tangential velocity has a discontinuity across the surface which increases with the distortion. For very flattened spheroids the form of the discontinuities is sketched in Fig. 9. Outside the surface the irrotational velocity is greatest at the equator, but inside it the greatest velocities in Hill's (1894) spheroidal vortex occur at a finite fraction of the radius. In the cylindrical polar coordinates (m, s) shown in Fig. 9, the velocity just inside the very flattened spheroid is approximately $\omega s = \Omega ms \propto ms$, where ω is the vorticity, and ms is a maximum where the eccentric angle is $\frac{1}{4}\pi$, i.e. at $1/\sqrt{2}$ of the distance from the axis to the equator. Consequently, the boundary layer around the surface contains velocity variations of the same order as the velocity itself, and its theory would be as complicated as that of the boundary layer on a solid body. It does not seem to have been developed, and so the shapes of drops can be accounted for only qualitatively.

Falling drops tend to be of the shape shown in Fig. 10, with the leading surface flatter than the trailing (Lane and Green, 1956; Winnikow and

---

FIG. 8. Drag coefficients plotted against Reynolds numbers for liquid drops. Experimental curves are drawn solid, theoretical curves dotted. Curves 1 and 2: Harper and Moore's theory (1968) without and with the logarithmic term. Curve 3: rigid spheres. Curve 4: Weber number $W_0 = 4$.

System	M_0	Dispersed phase	Continuous phase	Reference
A	5.1×10^{-11}	Carbon tetrachloride	Water	Thorsen *et al.* (1968)
B	2.2×10^{-10}	Ethylene bromide	Water	Thorsen *et al.* (1968)
C	3.1×10^{-11}	*o*-dichlorobenzene	Water	Thorsen *et al.* (1968)
D	9.1×10^{-11}	Ethyl bromide	Water	Thorsen *et al.* (1968)
E	2.55×10^{-8}	Water	Finol	Elzinga and Banchero (1961)
F	7.1×10^{-10}	Ethyl chloroacetate	Water	Licht and Narasimhamurty (1955)
G	1.9×10^{-10}	Methylene bromide	Water	Thorsen *et al.* (1968)
H	1.7×10^{-11}	Chlorobenzene	Water	Winnikow and Chao (1966)
K	5.8×10^{-11}	Bromobenzene	Water	Winnikow and Chao (1966)

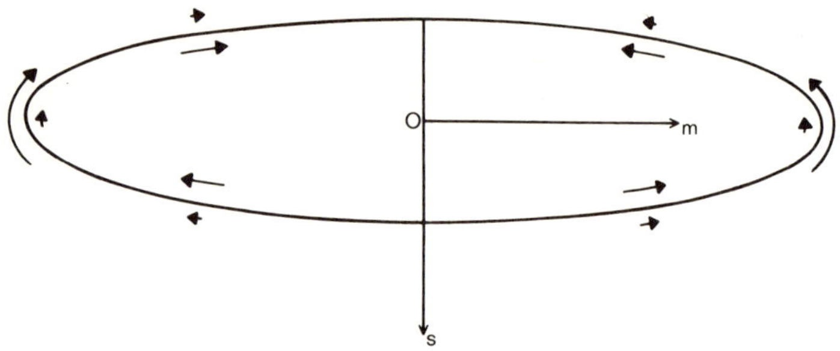

FIG. 9. Schematic indication of velocity variations around the surface of a falling spheroidal drop when very oblate. (Potential flow outside, Hill's spheroidal vortex inside.)

Chao, 1966). This is the opposite way round to bubbles. It seems that the high dynamic pressure near the front stagnation point A pushes it inwards, and surface tension is less impeded than it is for bubbles in its tendency to make the rear surface spherical because the density difference between inner and outer fluids is smaller in the experiments which have been reported,

Shoemaker and Marc de Chazal (1969) studied drops rising in high-M_0 fluids at high Weber numbers. They found skirts like those behind some bubbles (Section II, E, 3), and also reentrant "dimples" in the rear stagnation regions. So did Thomson and Newall (1885), for falling drops.

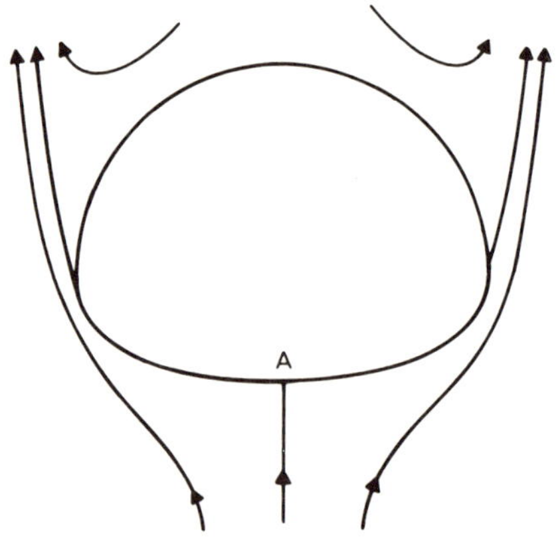

FIG. 10. A typical shape for a distorted falling drop, after Lane and Green (1956), with flow pattern sketched.

IV. Surface Activity

A. Introduction

It will be obvious to anyone familiar with the literature that the experimental data cited above as agreeing with theory are a very small fraction of those which have been published. Experimental liquids (especially water) are very easily contaminated with substances which lower the surface tension by an amount depending on the concentration. Fluid motions at the surface are then retarded, by the following mechanism. If the motion expands an element of area at the surface, a given amount of surfactant will have more surface to occupy (until diffusion restores equilibrium), the surface concentration will fall, and the surface tension will rise. This causes a tangential stress tending to drag fluid in along the surface towards the element of area, which opposes the original motion. Similarly contractions of elements of surface are also opposed.

B. Surface Concentrations

Although there are several different ways to define surface concentrations (see Defay *et al.*, 1966; Adam, 1968), most authors do not take the trouble to show that the same definition is being used in their equations for surface tension changes and conservation of mass of surfactant. This can be done by using Guggenheim's (1949) surface model and Landau and Lifshitz's (1959) theory of convective diffusion. The differences between the definitions are important for weakly surface-active solutes, such as ethyl alcohol in water.

We shall use the "continuum approximation," assuming quantities to vary continuously with position and ignoring molecular fluctuations, and we also assume that the relations between thermodynamical variables which hold at equilibrium remain good approximations for individual "infinitesimal" fluid elements, even when the system as a whole is in motion and its state varies from place to place. In this context, "infinitesimal" means "much smaller than macroscopic length-scales of variation, such as boundary-layer thicknesses, but much larger than the distances between molecules or their mean free paths." For a fuller discussion of this approximation, see Lighthill's Chapter I in Rosenhead (1963).

We follow Guggenheim (1949) in ignoring effects of curvature of the interface. These can be allowed for (see Defay *et al.*, 1966; Melrose, 1968), but they do not significantly modify the argument given below except for really minute bubbles or drops. The corrections become important at radii of curvature of about 1 μm (apart from the Young-Laplace pressure difference across the surface which may matter mechanically whenever the radius is less than 1 cm or thereabouts, but which has little effect on the surface thermodynamics at such large radii).

The Gibbs—Duhem equation connecting changes in surfactant concentration and surface tension may be derived as follows. Let us indicate the bulk phases as before by subscripts 0 and 1, and let phase i be a solution of surfactant (chemical component number 2) in a solvent (component number i). Let the mole fraction of surfactant be x_i, i.e. in a volume of phase i which contains k_i moles altogether, there are $k_i x_i$ moles of component 2 and $k_i(1-x_i)$ moles of component i. Throughout this section, $i=0$ or 1. To simplify the algebra we suppose that there are only these three components in the system, that none of them associates, dissociates, or reacts chemically with another, and that components 0 and 1 are mutually insoluble. (Otherwise there would be at least three different components in one or both phases.)

On each side of the interface and parallel to it, imagine planes drawn in the bulk phases, a distance l apart (Fig. 11). Suppose that l is large enough for conditions in each phase at its dividing plane to be uniform and unaffected by the presence of the other phase. Except in dilute ionic solutions l need only be a few molecular lengths; but l in an ionic solution must be several times the mean thickness of the double layer, which is $0.3c^{-1/2}$nm if c is the concentration of a dilute uni-univalent solute in mol dm^{-3} (see Parsons, 1954; J. T. Davies and Rideal, 1963). Ionic solutions, of course, require at least three chemical components to be considered in the same phase: positive and negative ions and bulk solvent. Their general theory is beyond the scope of this article.

The material between the two dividing planes (see Fig. 11) forms the

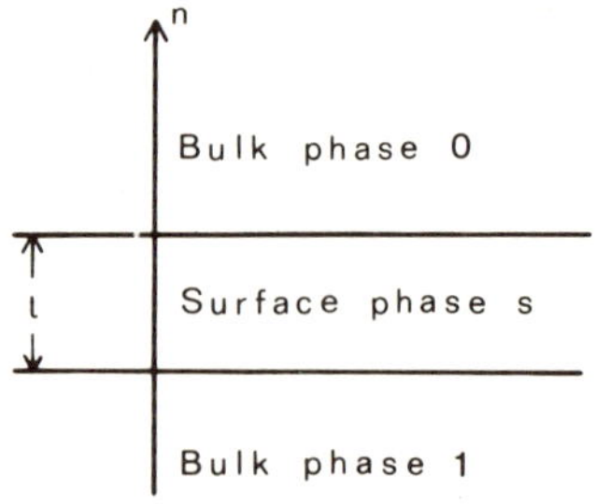

FIG. 11. Diagram of a fluid interface, showing the conventions for numbering phases and the direction of the normal. Bulk phase 0: component 2 dissolved in component 0. Surface "phase" s: all three components. Bulk phase 1: component 2 dissolved in component 1.

"surface phase" which we indicate by a subscript s. Unlike the bulk phases it is not even approximately homogeneous, and many of its properties depend on the value of l and the exact position within it of the true phase boundary. We shall be concerned in our final equations only with quantities independent of these geometrical parameters.

If the number of moles of component i ($i = 0$, 1, or 2) and the total entropy in area A of the surface phase are respectively n_{si} and S_s, let us define the surface concentration Γ_i and entropy per unit area s_s by

$$\Gamma_i = n_{si}/A, \tag{4.1}$$

$$s_s = S_s/A. \tag{4.2}$$

Guggenheim (1949) obtains the Gibbs-Duhem equation relating infinitesimal changes in intensive variables in the surface phase as

$$s_s\, dT - l\, dP + d\sigma + \sum_{i=0}^{2} \Gamma_i\, d\mu_i = 0, \tag{4.3}$$

in our notation, where T, p, σ, μ_i are the temperature, pressure, surface tension, and the chemical potential of component i, and where we imagine infinitesimal changes in these quantities to occur with the dividing planes fixed. The corresponding equations for the two bulk phases are (Guggenheim, 1949)

$$x_0\, d\mu_2 + (1 - x_0)\, d\mu_0 = 0, \tag{4.4}$$

$$x_1\, d\mu_2 + (1 - x_1)\, d\mu_1 = 0. \tag{4.5}$$

If l is so small that $l\, dp \ll d\sigma$, and if temperature changes are negligible, two terms in (4.3) disappear and the remaining ones give

$$-\left(\frac{\partial \sigma}{\partial \mu_2}\right)_T = \Gamma_2 - \Gamma_0 \frac{x_0}{1 - x_0} - \Gamma_1 \frac{x_1}{1 - x_1} = \Gamma, \tag{4.6}$$

say, when we substitute for $d\mu_1$ and $d\mu_0$ from (4.4) and (4.5), and use the fact that chemical potentials are uniform throughout the system. In a real system which may have large changes of μ_i over large distances, we must choose l so small that the μ_i are sensibly constant across the surface phase, having the same values at each dividing plane. Unless μ_i can have discontinuities, this requires only that l be much smaller than the diffusion boundary layer thickness. The question of possible discontinuities in μ_i is important and is taken up in Section IV, I below. For the moment we assume that they do not appear.

The quantity Γ defined by Eq. (4.6) is the "surface excess" of component 2, and is unaffected by movement of the imaginary dividing planes, unlike the three Γ_i. This can be seen either from the definitions of the Γ_i and of mole fractions, or from the invariance of the left-hand side of (4.6) under changes of position of the dividing planes. Physically, Γ is the amount of surfactant between the dividing planes, less the amount which would have been there if each bulk phase had continued unaltered right up to the plane of separation.

Further progress depends on knowledge of μ_i as a function of Γ or x_i. Let us suppose that the surfactant is dilute enough for it to be an ideal solution with $x_i \ll 1$ in each bulk phase and a "gaseous film" (see Adam, 1968) at the interface, i.e.

$$\begin{aligned}\mu_2 &= \bar{\mu}_{20} + R_g T \ln x_0 \\ &= \bar{\mu}_{21} + R_g T \ln x_1 \\ &= R_g T \ln(\Gamma/\bar{\Gamma}),\end{aligned} \tag{4.7}$$

where $\bar{\mu}_{20}$, $\bar{\mu}_{21}$, and $\bar{\Gamma}$ depend only on T, p, and the chemical natures of the three components, and R_g is the gas constant. Then $x_0 \propto x_1 \propto \Gamma$, and Eq. (4.6) gives

$$\Pi = \sigma_p - \sigma = R_g T\Gamma, \tag{4.8}$$

where σ_p is the interfacial tension between pure components 0 and 1, σ the tension when surface excess of component 2 is adsorbed on the interface, and Π is the "surface pressure."

C. Diffusion Boundary Conditions

Mole fractions are convenient units for chemistry, but if one has to deal also with dynamics and diffusion it is helpful to use mass fractions w_0, w_1 in the two phases. These are analogous to the mole fractions x_0, x_1, being the ratio of the mass of component 2 to the total mass of fluid in a given volume element. Landau and Lifshitz (1959) refer to a mass fraction as a concentration, c, but we follow the physical chemists' convention (McGlashan, 1968) of using this name and the symbol c_i for the amount of component 2 measured in moles per unit volume of phase i. If the molar masses of the three components are m_0, m_1, and m_2, and the total density of phase i is ρ_i, we have

$$\frac{m_2 c_i}{\rho_i} = w_i = \frac{m_2 x_i}{m_i(1 - x_i) + m_2 x_i} \doteqdot \frac{m_2 x_i}{m_i}, \tag{4.9}$$

where the last form on the right-hand side is an approximation for dilute solutions. In the bulk phase i, the mass flux of component 2 is $\rho_i w_i \mathbf{u}_i$ by convection with the fluid and, to a dilute solution approximation (Landau and Lifshitz, 1959),

$$\mathbf{j}_i = -\rho_i D_i \nabla w_i \doteqdot -m_2 D_i \nabla c_i \tag{4.10}$$

by diffusion, where $\mathbf{u}_i$ is the velocity vector (= momentum of unit mass), $\mathbf{j}_i$ the diffusion flux, and D_i the diffusivity. With these definitions of density and velocity, the usual equation of continuity and the Navier-Stokes equation hold in the bulk phase. Equation (4.10) for the diffusion flux holds, as Landau and Lifshitz explain, in the absence of thermal diffusion and barodiffusion, and then conservation of mass of solute gives

$$\partial c_i/\partial t = D_i \nabla^2 c_i, \tag{4.11}$$

if we can ignore the variations of ρ_i and D_i with concentration. Whether variations of ρ_i are important or not depends on the type of fluid motion under discussion. They usually do not matter for bubbles and drops, where they are much smaller than the density difference between the two bulk phases. They can, of course, be the cause of fluid motion in a system which would otherwise be in mechanical equilibrium, and then they must be taken into account.

We have not yet used the condition that mass of each chemical component must be conserved. Let the two dividing surfaces of Fig. 11 be fixed relative to the interface, so close to it that the tangential velocity $\mathbf{u}_s$ is effectively the same at all points between them. (This condition is usually less restrictive than the one that the chemical potentials be effectively uniform across the "surface phase.") Let the normal velocity components in the two phases at the dividing surfaces be u_{n0}, u_{n1}, both measured positively if in the direction of the vector from phase 1 to phase 0 in Fig. 11. As Levich (1962) pointed out, one cannot assume that the tangential mass flux $\mathbf{j}_{si}$ of component i is the same as its convective part $\Gamma_i m_i \mathbf{u}_s$, and we shall follow his course of taking the next simplest assumption. This is to suppose that $\mathbf{j}_{si} - \Gamma_i m_i \mathbf{u}_s$ is linearly dependent on the surface concentration gradients. Because any one of the Γ_i determines the other two if the temperature, pressure, and dividing surfaces are fixed, by the phase rule, we may use the surface gradient (Weatherburn, 1927; Scriven, 1960) of surface excess, $\nabla_s \Gamma$, to specify the surface concentration gradients. Then

$$\mathbf{j}_{si} = m_i(\Gamma_i \mathbf{u}_s - D_{si} \nabla_s \Gamma), \tag{4.12}$$

where the D_{si} are three constants of proportionality with the dimensions (L^2T^{-1}) of diffusivity. At most two of them can be independent, for

$$\sum_{i=0}^{2} \mathbf{j}_{si} = \left(\sum_{i=0}^{2} m_i \Gamma_i\right)\mathbf{u}_s = \left(\sum_{i=0}^{2} m_i \Gamma_i\right)\mathbf{u}_s - \left(\sum_{i=0}^{2} m_i D_{si}\right)\nabla_s \Gamma, \tag{4.13}$$

where the first part of (4.13) comes from the definition of total mass flux and the second from (4.12). Hence

$$\sum_{i=0}^{2} m_i\, D_{si} = 0. \tag{4.14}$$

By considering the total mass flux of the three components into an element of surface phase, we may now obtain

$$\frac{\partial \Gamma_0}{\partial t} + \operatorname{div}_s \mathbf{j}_{s0} = \frac{1}{m_0}\left(-\rho_0(1-w_0)u_{n0} - \rho_0\, D_0 \frac{\partial w_0}{\partial n}\right), \tag{4.15}$$

$$\frac{\partial \Gamma_1}{\partial t} + \operatorname{div}_s \mathbf{j}_{s1} = \frac{1}{m_1}\left(\rho_1(1-w_1)u_{n1} + \rho_1\, D_1 \frac{\partial w_1}{\partial n}\right), \tag{4.16}$$

$$\frac{\partial \Gamma_2}{\partial t} + \operatorname{div}_s \mathbf{j}_{s2} = \frac{1}{m_2}\left(-\rho_0 w_0 u_{n0} + \rho_1 w_1 u_{n1} + \rho_0\, D_0 \frac{\partial w_0}{\partial n} - \rho_1 D_1 \frac{\partial w_1}{\partial n}\right), \tag{4.17}$$

where $\operatorname{div}_s \mathbf{j}_{si}$ denotes the surface divergence of the vector $\mathbf{j}_{si}$ (Weatherburn, 1927); i.e. Scriven's (1960) $\nabla_s \cdot \mathbf{j}_{si}$. Because the surface tension depends on Γ, we must take a linear combination of the above equations of the form

$$(4.17) - \frac{x_0}{1-x_0}(4.15) - \frac{x_1}{1-x_1}(4.16),$$

and we obtain

$$\begin{aligned}\frac{\partial \Gamma}{\partial t} &+ \operatorname{div}_s(\Gamma \mathbf{u}_s) + \sum_{i=0}^{1} \frac{\Gamma_i m_i}{m_2}\left(\frac{\partial}{\partial t} + \mathbf{u}_s \cdot \nabla\right)\left(\frac{w_i}{1-w_i}\right) \\ &= -\sum_{i=0}^{1}\left(\frac{1}{m_2(1-x_i)} \operatorname{div}_s(D_{si} m_i \nabla_s \Gamma) + \frac{(-1)^i \rho_i D_i}{m_2(1-w_i)} \frac{\partial w_i}{\partial n}\right).\end{aligned} \tag{4.18}$$

This equation simplifies, if we use the bulk diffusion equation (4.13), the dilute-solution approximations $w_i \ll 1$, $x_i \ll 1$, D_i constant, and D_{si} constant, and the definition $c_i = w_i \rho_i / m_2$, to

$$\begin{aligned}\frac{\partial \Gamma}{\partial t} &+ \operatorname{div}_s(\Gamma \mathbf{u}_s) + \sum_{i=0}^{1} \frac{\Gamma_i m_i}{m_2}\left(D_i \nabla^2 w_i - u_{ni} \frac{\partial w_i}{\partial n}\right) \\ &= D_{s2} \nabla_s{}^2 \Gamma + \sum_{i=0}^{1} (-1)^i D_i \frac{\partial c_i}{\partial n}.\end{aligned} \tag{4.19}$$

If we assume that l, the thickness of the interfacial region, is much smaller than the diffusion boundary layer thickness, then the terms in Γ_i in (4.19) are negligible. Here the boundary layer thickness is best thought of as the shorter of the distances in the n direction over which $\partial w_0/\partial n$ and $\partial w_1/\partial n$ change by significant fractions of themselves. If l is not this small, then using it as a thickness of a surface phase is unhelpful. We shall therefore neglect those terms in Γ_i, and recover Levich's (1962) surface condition

$$\frac{\partial \Gamma}{\partial t} + \mathrm{div}_s(\Gamma \mathbf{u}_s) = D_s \nabla_s{}^2 \Gamma + D_0 \frac{\partial c_0}{\partial n} - D_1 \frac{\partial c_1}{\partial n} \tag{4.20}$$

on writing $D_{s2} = D_s$, but we have used a definition of Γ and hence of D_s more explicit than his.

Equation (4.20) now appears as one of the two boundary conditions interconnecting the concentration and velocity fields at an interface between surfactant solutions. The other condition derives from the fact that for small l there is so little mass in the surface phase that the forces on it can be taken to be in equilibrium. The normal component gives the Young-Laplace relation between normal stresses, surface tension, and curvature (see Section II,C,1 and Eq. 2.44), and the tangential component gives the condition we seek, in the form (Scriven, 1960)

$$\nabla_s \sigma = -\nabla_s \Pi = p_{ns1} - p_{ns0} \tag{4.21}$$

where p_{nsi} is the shear stress at the surface in phase i. The relation between (4.21) and concentrations is clear from Eqs. (4.7)–(4.9), which we may use to write

$$\nabla_s \sigma = -(R_g T \Gamma / c_i) \nabla_s c_i = -R_g T \nabla_s \Gamma, \quad \text{for} \quad i = 0, 1, \tag{4.22}$$

for ideal dilute solutions.

The thermodynamics of ideal solutions gives no information about the distribution coefficients c_0/c_1 and Γ/c_i except that they are constants relating the concentrations of the solute in the two bulk phases and the surface. Their values are determined by the chemical natures of the three components. We define "adsorption depths" h_0, h_1 by

$$h_i = \Gamma / c_i, \quad \text{for} \quad i = 0, 1; \tag{4.23}$$

they have the dimensions of length because Γ is an amount of substance per unit area and c_i an amount per unit volume. The depth h_i of phase i obviously contains the same amount of dissolved solute (component 2) as is adsorbed on its surface. Numerical values of h_i cover a very wide range. In water, at an air–water interface, h_i is of the order of a molecular length for a weakly adsorbed solute like ethyl alcohol C_2H_5OH [4 nm according to Adam (1968), 7 nm according to Bakker *et al.* (1966)], but for strongly adsorbed substances it is much higher, especially if they are sparingly soluble. [1 μm for sodium dodecyl sulphate, $C_{12}H_{25}SO_4Na$ (Durham, 1961; Rubin and Jorne, 1969), 0.3 mm for dodecanol, $C_{12}H_{25}OH$ (Shinoda and Nakanishi, 1963).] By Traube's rule (see Adam, 1968), h_i increases by a factor of about 3 for each additional CH_2 group in a homologous series of long chain organic compounds. But the values are difficult to determine; discrepancies by nearly a factor of 2 occur for the same substance, as in the two values given above for ethyl alcohol, and the "Traube factor" is sometimes nearer 4 than 3, as in Fig. 7.1 of Defay *et al.* (1966), reporting work of Hommelen on the *n*-aliphatic alcohols $C_mH_{2m+1}OH$ for $m = 6, 7, 8, 9$, and 10. Hommelen's solutions, however, were too concentrated to behave as if ideal.

If more than one surfactant is present, but all of them are very dilute, Eq. (4.20) still holds for each unless the rate of chemical change of one into another is comparable with the rate of transfer along or into the surface. Equation (4.22) becomes

$$\begin{aligned} \nabla_s \sigma &= -R_g T \sum_{j=2}^{n} \frac{\Gamma_j'}{c_{ij}} \nabla_s c_{ij} \\ &= -R_g T \sum_{j=2}^{n} \nabla_s \Gamma_j', \quad \text{for} \quad i = 0, 1, \end{aligned} \tag{4.24}$$

where c_{ij} is the concentration of component j in bulk phase i, the bulk solutes in phases 0, 1 are components 0, 1 as before, and the surface-active solutes are components $2, 3, \ldots, n$. Γ_j' is the generalization of Γ, obtained in the following way. Let x_{ij} be the mole fraction of component j in phase i, so that

$$\sum_{j=0}^{n} x_{ij} = 0 \quad \text{for} \quad i = 0, 1, \tag{4.25}$$

and in dilute solutions $x_{00} \doteqdot 1$, and $x_{11} \doteqdot 1$. The Gibbs-Duhem equations for the surface and the bulk phases are, at constant temperature and pressure,

$$-d\sigma = \sum_{j=0}^{n} \Gamma_j \, d\mu_j, \tag{4.26}$$

$$0 = \sum_{j=0}^{n} x_{ij} \, d\mu_j, \quad \text{for} \quad i = 0, 1, \tag{4.27}$$

where μ_j is the chemical potential of component j. On eliminating $d\mu_0$ and $d\mu_1$ from Eqs. (4.26) and (4.27) we obtain an equation giving $-d\sigma$ in terms of $d\mu_2, d\mu_3, \ldots, d\mu_n$, and define Γ_j' to be the coefficient of $d\mu_j$ in it. In very dilute solutions

$$\begin{aligned} \Gamma_j' &= -\partial\sigma/\partial\mu_j \\ &\doteqdot \Gamma_j - x_{0j}\Gamma_0 - x_{1j}\Gamma_1, \quad \text{for} \quad j, k = 2, 3, \ldots, n, j \neq k. \end{aligned} \tag{4.28}$$

In an ionic solution this type of analysis is necessary because there must be at least the two oppositely charged ions present together with the solute. Then one has also to use the condition of electrical neutrality in the bulk phases to show that the effective diffusivity D_e is

$$D_e = \frac{D^+ D^-}{z^+ D^+ + z^- D^-}, \tag{4.29}$$

when there are two types of ion with electrovalencies z^+, z^- and diffusivities D^+, D^- (see Deryagin *et al.*, 1959; Levich, 1962). In addition the surface layer as a whole must be electrically neutral although the actual surface is usually charged, so that the immediately subjacent fluid has an excess of ions of the opposite charge. We shall not discuss these "double layers" here (see J. T. Davies and Rideal, 1963; Parsons, 1954), nor the sort of complication which arises with acetic acid at the benzene–water interface, when $(CH_3COOH)_2$ and CH_3COOH molecules occur in the benzene, and CH_3COO^- and H_3O^+ as well in the water, and reversible reactions occur between the various species. Such behavior is common among organic acids, at air–water or benzene–water interfaces (Glasstone, 1953).

Heat can also be considered as a surfactant, because changes in temperature, like changes in chemical potential, affect the surface tension and give rise to diffusion. One can draw up a list of quantities analogous to one another in the theory (see Harper *et al.*, 1967): D_i to the heat diffusivity $\kappa_i = K_i/\rho_i c_{pi}$, where K_i is the heat conductivity and c_{pi} the specific heat at constant pressure of phase i, μ_j to $R_g T$, Γ to $-(1/R_g)(\partial\sigma/\partial T)_{\mu_j}$, h_i to $-(1/\rho c_{pi})(\partial\sigma/\partial T)_{\mu_j}$, where μ_j is the chemical potential. Because its "h_i" is very small (0.037 nm for water at room temperature) the surface effects of heat can usually be ignored in practice unless, of course, they cause the motion, as in some Bénard cells (Pearson, 1958; Scriven and Sternling, 1960, 1964), or are artificially increased as in the work of Young *et al.* (1959) on bubbles held fixed against gravity in a vertical temperature gradient.

Dimensional analysis takes a form in this subject which may be a little unfamiliar. Besides the usual mass M, length L, and time T of mechanics as dimensionally independent variables, we have temperature θ and amount of substance Q (McGlashan, 1968). The unit for Q in the metric system is the mole. The analogies in the previous paragraph have been altered slightly from those of Harper *et al.* (1967) to make analogous quantities have the same dimensions. We collect the quantities used in this chapter with their dimensions, in Table 4.

TABLE 4

DIMENSIONS OF PHYSICOCHEMICAL QUANTITIES

c_i, c_{ij}	concentration	QL^{-3}
c_p	specific heat	$L^2T^{-2}\theta^{-1}$
D_s, D_{si}, D_i	diffusivity	L^2T^{-1}
h_i	adsorption depth	L
$\mathbf{j}_i$	volume mass flux	$ML^{-2}T^{-1}$
$\mathbf{j}_{si}$	surface mass flux	$ML^{-1}T^{-1}$
K_i	heat conductivity	$MLT^{-3}\theta^{-1}$
m_i	molar mass	MQ^{-1}
n_{si}	amount of i in surface phase	Q
R_g	gas constant	$ML^2T^{-2}Q^{-1}\theta^{-1}$
S_s	entropy	$ML^2T^{-2}\theta^{-1}$
s_s	entropy per unit area	$MT^{-2}\theta^{-1}$
T	temperature	θ
w_i	mass fraction	1
x_i	mole fraction	1
Γ, Γ_i'	surface excess	QL^{-2}
Γ_i	surface phase concentration	QL^{-2}
κ_i	heat diffusivity	L^2T^{-1}
μ_i	chemical potential	$ML^2T^{-2}Q^{-1}$
σ	surface tension	MT^{-2}
Π	surface pressure	MT^{-2}

D. A Drop or Bubble Moving at Low Péclet Number

To illustrate the interaction of fluid motion and solute diffusion, let us consider the case of a steadily moving drop which is small enough to be considered spherical and whose Péclet numbers Ud/D_i, Ud/D_s are much smaller than one. Because D_i is usually much less than ν_i, the Reynolds numbers will be small too. Bubbles obey the same theory if η_1 is neglected in the results. We generalize slightly the theories of Frumkin and Levich (see Levich, 1962); Harper *et al.* (1967); and Kenning (1969).

The equation of motion is $D^4\psi = 0$ in each phase (Sections II,C,1 and III,B,1), and that of diffusion is $\nabla^2 c_i = 0$ for $i = 0, 1$, if the Péclet numbers are small. The boundary conditions are Eqs. (4.20) to (4.23), together with continuity of velocity at the interface, finiteness of velocity and c_1 at the origin, and uniform limiting values U and c_∞ for velocity and c_0 at infinity. It is convenient to define a fictitious "surface pressure" throughout the space as

$$\Pi = R_g T h_i c_i, \tag{4.30}$$

where h_i takes its interfacial value everywhere in phase i. This definition makes Π a harmonic function in each phase and continuous across the interface $r = a$. At the interface it keeps its previous value and meaning, defined in (4.8). The value of Π at infinity, Π_∞, is a convenient measure of the strength of the solution, being given by

$$\Pi_\infty = R_g T h_0 c_\infty, \tag{4.31}$$

the amount by which the equilibrium interfacial tension between phases 0 and 1 is lowered by the presence of the surfactant. Then

$$\Pi = \Pi_\infty + \sum_{n=0}^{\infty} \varepsilon_n \frac{r^n}{a^n} P_n(\mu) \tag{4.32}$$

inside the drop and

$$\Pi = \Pi_\infty + \sum_{n=0}^{\infty} \varepsilon_n \frac{a^{n+1}}{r^{n+1}} P_n(\mu) \tag{4.33}$$

outside it, for some set of constants ε_n, where $\mu = \cos\theta$. The surface stress condition (4.21) can be written

$$\varepsilon_1 = \tfrac{3}{2} U [(3 - 2\alpha_2)\eta_1 - 2(\alpha_2 - 1)\eta_0], \tag{4.34}$$

$$\varepsilon_n = -(4n + 2) U \alpha_{n+1}(\eta_0 + \eta_1)/n(n+1) \quad \text{for} \quad n \geq 2, \tag{4.35}$$

if we use Eqs. (3.6) and (3.7), and equate coefficients at $\mathscr{I}_n(\mu)$. No information comes from this source about ε_0, the difference between the surface mean value of Π and Π_∞. The surface mass balance equation (4.20) becomes

$$U \frac{d}{d\mu}\left\{\left[\Pi_\infty + \sum_0^\infty \varepsilon_n P_n(\mu)\right]\left[(3 - 2\alpha_2)\mathscr{I}_2(\mu) - 2\sum_3^\infty \alpha_n \mathscr{I}_n(\mu)\right]\right\}$$
$$= \sum_0^\infty \varepsilon_n P_n(\mu)\left\{n(n+1)\frac{D_s}{a} + (n+1)\frac{D_0}{h_0} + n\frac{D_1}{h_1}\right\}. \tag{4.36}$$

By integrating this equation from $\mu = -1$ to $\mu = 1$ and using the definition of $\mathscr{I}_n(\mu)$ (Eq. 2.11), it can be shown that $\varepsilon_0 = 0$. We do not attempt to equate any other coefficients of $P_n(\mu)$ in the general case. It would require the expansion of $P_n(\mu)\mathscr{I}_m(\mu)$ in Legendre polynomials, lead to an infinite set of nonlinear equations to solve for the α_i, and be of spurious physical generality anyway, as we shall see in the discussion following Eq. (4.40). It will appear that in cases likely to arise in real fluids the surface pressure Π is only slightly perturbed from its mean value Π_∞. With that assumption, to be checked later, we may ignore the term $\sum e_n P_n(\mu)$ on the left-hand side of Eq. (4.36) and obtain $\alpha_{n+1} = \varepsilon_n = 0$ for $n \geq 2$, while

$$\begin{aligned}\alpha_2 &= R_0 C_{D0}/16 \\ &= (2\eta_0 + 3\eta_1 + \Pi_\infty/v_D)/(2\eta_0 + 2\eta_1 + 2\Pi_\infty/3v_D),\end{aligned} \tag{4.37}$$

where the "diffusion velocity" v_D is defined by

$$v_D = \frac{D_s}{a} + \frac{D_0}{h_0} + \frac{D_1}{2h_1}. \tag{4.38}$$

It is additive over the three routes for diffusion—phase 0, phase 1, and surface—because a given inequality of surface concentration is "smoothed out" by the three "diffusion currents" in parallel. The surface velocity is $v_0 \sin\theta$, where

$$\begin{aligned} v_0 &= U(\tfrac{3}{2} - \alpha_2) \\ &= U\eta_0/(2\eta_0 + 2\eta_1 + 2\Pi_\infty/3v_D), \end{aligned} \tag{4.39}$$

and the surface pressure is $\Pi_\infty + \varepsilon_1 \cos\theta$, where

$$\begin{aligned} \varepsilon_1/\Pi_\infty &= -v_0/v_D \\ &= -U\eta_0/v_D(2\eta_0 + 2\eta_1 + 2\Pi_\infty/3v_D). \end{aligned} \tag{4.40}$$

Equation (4.40) shows that for the theory to be valid it is sufficient to require that the two Reynolds and three Péclet numbers, Ud/ν_i, Ud/D_i, and Ud/D_s, be all much less than one, for $2v_0 < U$ and $v_D > 2D_s/d$, and so $|\varepsilon_1/\Pi_\infty| < Ud/4D_s \ll 1$, by hypothesis. Wasserman and Slattery (1969) present a theory in which ε_1/Π_∞ is not small but the Reynolds numbers and two of the Péclet numbers (Ud/D_i) are, and they ignore surface diffusion. That part of their work seems unlikely to be a useful generalization unless $D_s \ll D_0$, a supposition for which there is no evidence yet.

The dynamical effect of the surfactant on the drag and the surface velocity is effectively to increase the viscosity inside the drop from η_1 to $\eta_1 + \Pi_\infty/3v_D$ and hence bring the motion nearer to that of a solid sphere. Clearly the influence is greater in more concentrated solutions (higher Π_∞), but the theory becomes invalid if the concentration is too high. The surface pressure varies from a minimum of $\Pi_\infty + \varepsilon_1$ at the front stagnation point to a maximum of $\Pi_\infty - \varepsilon_1$ at the rear, at least if $\Pi_\infty > 0$ and the surfactant is positively adsorbed. Some solutes raise the surface tension instead of lowering it, in which case Π_∞, Γ, h_i, and v_D are all negative. If so, the solute is never strongly surface-active (see Adam, 1968, Section III.7), and then $|h_i| \ll a$ for all practical sizes of drops, and surface diffusion can be neglected.

The implications of the theory are easiest to understand for bubbles, where $\eta_1 \ll \eta_0$ and $D_1/h_1 \ll D_0 h_0$ (unless the surfactant is volatile). Then $\Pi_\infty/\eta_0 v_D$ reduces for strong surfactants or small bubbles ($h_0 \gg a$) to $\Pi_\infty a/\eta_0 D_s$, and for weak surfactants or large bubbles ($h_0 \ll a$) to $\Pi_\infty h_0/\eta_0 D_0$. For a moderately surface-active substance in water, with $h_0 \doteqdot 1$ μm (for example, the dodecyl sulphate ion, or hexanol),

$D_0 \doteqdot 5 \times 10^{-10}$ m^{2}sec$^{-1} = 5 \times 10^{-6}$ cm^2 sec^{-1}, and so $\Pi_\infty h_0/\eta_0 D_0 = 1$ if $\Pi_\infty = 0.5$ μN m$^{-1} = 5 \times 10^{-4}$ dyn cm^{-1}. The surface concentration is then 2×10^{-10} mol m^{-2} (8×10^4 square Ångström units per molecule) and the bulk concentration 2×10^{-4} mol m^{-3}.

Bubble behavior might seem from these figures to provide a good means of experimental research into surface properties of solutions too dilute to measure by other methods, but there is a difficulty. We have assumed the Péclet number to be small. The velocity U is at least its Stokes law value $2ga^2/9\nu$, and so $Ud/D_0 < 1$ only if $d < 10$ μm, for the example above. We can therefore apply the present theory in water only for very small bubbles: 10 μm = 0.01 mm. The situation is not much better in most other liquids, for $Ud/D_0 \geq gd^3/18D_0\nu_0$, and the Stokes-Einstein theory of diffusion (see Bird *et al.*, 1960) gives

$$D_0\nu_0 = kT/6\pi\rho_0 a_m, \tag{4.41}$$

where k is Boltzmann's constant and a_m is the radius of a surfactant molecule if it is spherical, or a distance of the order of the molecule's size if not. For a given surfactant, then, $Ud/D_0 \propto d^3\rho_0/T$ if the bubble size and the solute are varied. Larger bubbles can therefore satisfy the conditions of the theory for small Péclet numbers only if the temperature is high, the density low, or the Stokes-Einstein law is not obeyed. There are viscous liquids with that last property, such as Redfield and Houghton's (1965) dextrose solutions in which the diffusivity of carbon dioxide was anomalously high, but carbon dioxide, of course, is not very surface-active.

For drops the conditions are slightly less restrictive, because $Ud/D_0 < |\rho_0 - \rho_1| gd^3/12D_0\nu_0\rho$ if we use the fact that

$$U < |\rho_0 - \rho_1| ga^2/3\nu_0\rho_0,$$

which is the Rybczynski-Hadamard result for $\eta_1 = 0$. The gain is, however, not great. To multiply the permissible value of d by 10, $|\rho_0 - \rho_1| < 10^{-3}\rho_0$; experiments with such small density differences would need elaborate precautions against convection currents.

E. A Bubble at Low Reynolds and High Péclet Number

We have seen that bubbles visible to the naked eye are unlikely to obey the requirement of Section IV,D that the Péclet number be small, and so the next simplest case, of low Reynolds and high Péclet numbers, has received a good deal of attention (Deryagin *et al.*, 1959, 1960; Derjaguin and Dukhin, 1961; Dukhin and Deryagin, 1961; Levich, 1962; Griffith, 1962; Dukhin and Buikov, 1965; Lochiel, 1965; Davis and Acrivos, 1966;

Newman, 1967; Wasserman and Slattery, 1969). Unfortunately some of this work uses unjustified assumptions, and some authors neglect singularities which predict infinite surface excesses at rear stagnation points. Very little has been done for Péclet numbers of order unity except by Wasserman and Slattery (1969), whose numerical method becomes untrustworthy at high Péclet numbers because of those same singularities. All this work ignores transport of surfactant through the interior. To allow for it would involve the difficulties, caused by recirculation of fluid inside the bubble, that we encountered for high Reynolds numbers in Section III,C,2.

For a spherical bubble in steady flow the surface condition (4.20) becomes

$$a\frac{\partial}{\partial\theta}(\Pi u_\theta \sin\theta) = D_s \frac{\partial}{\partial\theta}\left(\sin\theta\frac{\partial\Pi}{\partial\theta}\right) + \frac{a^2 D_0 \sin\theta}{h_0}\frac{\partial\Pi}{\partial r}, \tag{4.42}$$

in terms of the surface pressure Π, defined on the surface by (4.8) and elsewhere by (4.30).

1. *Slightly Contaminated Surface, Weak Surfactant*

Equation (4.42) involves Π and u_θ, both of which are unknown in general. This makes the general theory very difficult, and so it is worthwhile to seek special cases of physical interest which are simpler. The first is that in which u_θ is very near the value $v_0 \sin\theta$ given by the Rybczynski-Hadamard theory for a free surface, i.e. the surfactant has had only a slight effect on the motion of the bubble. Equation (4.42) is then linear in its remaining unknown variable Π, but is still complicated. Suppose that the surface diffusion term in (4.42) can be neglected, and that $\Pi \to \Pi_\infty$ at a large distance from the bubble. Then, as the Péclet number is high, analysis like that of Section II,C,2 gives

$$4\,\partial\Pi/\partial x = \partial^2\Pi/\partial y^2, \tag{4.43}$$

where $x = \frac{1}{4}(1-\cos\theta)^2(2+\cos\theta)$ as before, and

$$y = \frac{(r-a)\sin^2\theta}{d}\left(\frac{3v_0 d}{8D_0}\right)^{1/2}. \tag{4.44}$$

In (x, y) coordinates the boundary condition (4.42) becomes

$$\frac{\partial}{\partial x}(\Pi\sin^2\theta) = \frac{3}{2P_{vs}}\frac{\partial}{\partial x}\left(\sin^4\theta\frac{\partial\Pi}{\partial x}\right) + \frac{\delta_0}{h_0}\frac{\partial\Pi}{\partial y}, \tag{4.45}$$

where $P_{vs} = v_0 d/D_s$ is the Péclet number based on surface velocity and surface diffusivity, and δ_0, a convenient measure of the diffusion boundary layer thickness, is defined by

$$\delta_0{}^2 = D_0 d/6v_0 = d^2/6P_{v0}, \tag{4.46}$$

where $P_{v0} = v_0 d/D_0$.

Equation (4.45), with the conditions $\Pi \to \Pi_\infty$ as $y \to \infty$, $\Pi = \Pi_\infty$ at $x = 0$, $y > 0$, Π finite at $x = y = 0$, uniquely specifies the solution of (4.43). If x is small, $\sin^4 \theta \doteqdot 4(\frac{1}{3}x)^{1/2}$, and on making that approximation we find that

$$\Pi \doteqdot \Pi_\infty - (\Pi_\infty - \Pi_0)\mathrm{erfc}(y/x^{1/2}), \tag{4.47}$$

where the surface pressure Π_0 at the front stagnation point obeys

$$\Pi_0 = \Pi_\infty \delta_0/[\delta_0 + h_0(\pi/3)^{1/2}]. \tag{4.48}$$

This simple calculation shows that Π_0/Π_∞ is nearly zero if $\delta_0 \ll h_0$, and is nearly one if $\delta_0 \gg h_0$. These two cases correspond respectively to there being much less and much more surfactant dissolved in the boundary layer than adsorbed at the surface. For brevity we shall refer to the surfactant in these two cases as being strong and weak respectively.

The first of the two cases to be described in detail was that of a weak surfactant, by Frumkin and Levich in 1949 (see Levich, 1962). They made a rather crude approximation to $\partial\Pi/\partial y$, and Derjaguin and Dukhin (1961) improved it. But even they do not deal adequately with the singularities which appear at the rear stagnation point.

For a weak surfactant ($\delta_0 \gg h_0$) in very dilute solution, we may take Π everywhere around the bubble to be close to Π_∞. The neglect of surface diffusion in (4.45) is easily justified, and the equation reduces to

$$\frac{\partial\Pi}{\partial y} = \frac{h_0 \Pi_\infty}{\delta_0} \frac{\partial}{\partial x}(\sin^2 \theta) = \frac{h_0 \Pi_\infty}{4\delta_0} \frac{\partial^2 N(x, 0)}{\partial y^2}, \tag{4.49}$$

approximately, and the solution to

$$\Pi = \Pi_\infty\left(1 + \frac{h_0}{4\delta_0} \frac{\partial N(x, y)}{\partial y}\right). \tag{4.50}$$

In these equations $N(x, y)$ is the solution of the diffusion equation (4.43) with initial condition $N(0, y) = 0$ and boundary condition $N(x, 0) = \sin^2 \theta(x) = f(x)$, say, on $y = 0$, where $\theta(x)$ is the inverse function to $x(\theta)$. That is (Dukhin, 1966),

$$N(x, y) = \int_0^x \frac{df(t)}{dt} \,\mathrm{erfc}\, \frac{y}{(x - t)^{1/2}}\, dt. \tag{4.51}$$

From Eq. (4.51) we find (see Harper, 1972) that as $\mu = \cos\theta \to -1$, $\partial N/\partial y \to \infty$ on the surface $y = 0$, and that $\partial N/\partial y$ is asymptotic to an exact solution of the full diffusion equation near the rear stagnation point but it does not satisfy all the boundary conditions there. The form of the singularity is that of a line sink, with $\partial\Pi/\partial m \to \text{constant}/m$ as distance m from the axis of symmetry tends to zero. One must therefore add to the expression for Π a line-source solution of the full diffusion equation, of the correct strength to cancel that singularity. The correction is significant only in a small neighborhood $\pi - \theta = O(P_{v0}^{-1/2})$ of the stagnation point, in which $\Pi - \Pi_\infty$ is of order $\Pi_\infty h_0 \ln P_{v0}/\delta_0$ instead of $\Pi_\infty h_0/\delta$, its order of magnitude at the front stagnation point and in most of the boundary layer.

The drag coefficient is hardly affected by the correction, because the integral from which it is calculated converges even for the singular expression (4.50). To calculate the drag, the most convenient method is to use Eqs. (2.12), (2.13), and (4.21) to prove that

$$C_{D0} = \frac{16}{R_0}\left(1 + \frac{1}{2U\eta_0}\int_{-1}^{1} \mu\sigma(\mu)\,d\mu\right) = \frac{16}{R_0}\left(1 - \frac{1}{2U\eta_0}\int_{-1}^{1} \mu\Pi(\mu)\,d\mu\right); \tag{4.52}$$

in this equation $\sigma(\mu)$, $\Pi(\mu)$ are the values of the surface tension and surface pressure where $\cos\theta = \mu$. Hence

$$C_{D0} = \frac{16}{R_0}\left(1 + 0.608\,\frac{h_0\Pi_\infty}{\eta_0 D_0 P_0^{1/2}}\right), \tag{4.53}$$

where $P_0 = Ud/D_0 = 2P_{v0}$.

2. *Slightly Contaminated Surface, Strong Surfactant*

The quantitative theory for a strong surfactant ($h_0 \gg \delta_0$) was first given by Deryagin *et al.* (1960), but their treatment of the singularity at the rear stagnation point is inadequate. The error is worse than in the previous case. We have already seen in Eq. (4.48) that the surface pressure Π_0 at the front stagnation point is very much smaller than Π_∞ if surface diffusion is neglected, being $O(\Pi_\infty\delta_0/h_0)$.

If we assume that $\Pi_0 \ll \Pi_\infty$ even when surface diffusion is allowed for, there will be some part of the bubble's surface over which $\Pi \ll \Pi_\infty$, and then the normal diffusive flux to the surface will be approximately the same as if Π were put equal to zero. That is,

$$\frac{\partial\Pi}{\partial y} \doteqdot \frac{2\Pi_\infty}{(\pi x)^{1/2}}\exp\left(-\frac{y^2}{x}\right) = \frac{2\Pi_\infty}{(\pi x)^{1/2}} \quad \text{at} \quad y = 0, \tag{4.54}$$

from Eq. (4.43). Equation (4.54) can be rewritten in terms of $\partial\Pi/\partial r$, and the result substituted into (4.42) and integrated, to obtain the following differential equation for Π on the bubble surface:

$$\Pi + \frac{D_s}{av_0}\frac{\partial\Pi}{\partial\mu} = \frac{2\Pi_\infty\delta_0}{\pi^{1/2}h_0}\frac{(2+\mu)^{1/2}}{(1+\mu)}. \tag{4.55}$$

From this equation it appears that Π is of order $\Pi_\infty\delta_0/h_0$ not only at the front stagnation point but over most of the bubble surface, and that surface diffusion may be neglected over most of the surface, because D_s/av_0 is a small parameter. Near the rear stagnation point, however, these estimates fail, because $\varepsilon = 1 + \mu \sim \frac{1}{2}\varphi^2 \to 0$, where $\varphi = \pi - \theta$. Hence Π/Π_∞ does not remain small all the way round the bubble.

Harper (1972) has described how to resolve the singularity, but the complicated details will not be given here. He finds a first approximation to C_{D0} in the form of

$$C_{D0} = \frac{16}{R_0}\left(1 + \frac{\Pi_\infty}{\sqrt{3}U\mu}\right), \tag{4.56}$$

which is independent of D_s, although the distribution of surfactant at distances of order $aP_{vs}^{-1/2}$ from the rear stagnation point is not. At that point

$$\Pi = \Pi_\infty P_{vs}/\sqrt{3}. \tag{4.57}$$

For comparisons with experiments, it is easier to give C_{D0} for a given liquid and surfactant concentration in the form

$$C_{D0} = (16/R_0)(1 + Ad^n), \tag{4.58}$$

where the correction term Ad^n is assumed much smaller than unity, A is a constant, $n = -\frac{3}{2}$ for a weak surfactant, from (4.53), and $n = -2$ for a strong one, from (4.56).

F. Drops and High Reynolds Numbers

If the surfactant is soluble only in the continuous phase, all the theory in Section IV,E is applicable to drops, except that v_0 is now slightly perturbed from $\frac{1}{2}U\eta_0/(\eta_0 + \eta_1)$ instead of $\frac{1}{2}U$ (Eq. 3.9), and C_{D0} is given by a formula more elaborate than (4.52), because Eqs. (3.6) and (3.7) for the internal velocity and shear stress must be taken into account. The result is

$$C_{D0} = \frac{8}{R_0(\eta_0 + \eta_1)}\left(2\eta_0 + 3\eta_1 - \frac{1}{U}\int_{-1}^{1}\mu\Pi(\mu)\,d\mu\right), \tag{4.59}$$

which generalizes Eq. (4.52), and for a weak surfactant

$$C_{D0} = \frac{8}{R_0(\eta_0 + \eta_1)} \left[2\eta_0 + 3\eta_1 + 0.608 \frac{h_0 \Pi_\infty}{D_0 P_0^{1/2}} \left(\frac{\eta_0}{\eta_0 + \eta_1} \right)^{1/2} \right], \tag{4.60}$$

to the first order.

If the Reynolds number R_0 of a bubble is much larger than one, but not large enough for the bubble to be significantly distorted, it can be shown (Harper, 1972) that the analysis is again affected only in the value of v_0 and the final calculation of C_{D0}, which is now

$$C_{D0} = \frac{48}{R_0} \left(1 - \frac{1}{2U\eta_0} \int_{-1}^{1} \mu \Pi(\mu) \, d\mu \right) = \frac{48}{R_0} \left(1 + \frac{1.053 h_0 \Pi_\infty}{\eta_0 D_0 P_0^{1/2}} \right) \tag{4.61}$$

for a weak surfactant. In this case the perturbations reach larger fractions of the basic flow than for small R_0, and there is not much difference in aqueous solutions between the values of Π_∞ which are too small for C_{D0} to be noticeably affected and too large for the small perturbation theory to be applicable. The situation is worse for a strong surfactant, and the value of C_{D0} does not seem worth giving.

Unfortunately no theory seems to be available for finite perturbations at high Reynolds numbers. Proportional increases in drag coefficient can occur which are greater than at low Reynolds numbers, because the ratio of C_{D0} values for spheres with rigid and free surfaces increases with R_0 (see Fig. 1). For distorted bubbles the effect of surfactants diminishes, until it can be neglected for spherical caps, whose shape and drag can be explained fairly well without considering surface tension at all (see Section II,E).

To date, all the theory described in this chapter has been confirmed only qualitatively and semiquantitatively by experiments (Levich, 1962). Surfactant solutions with accurately known concentrations are difficult to prepare, and even more difficult to keep in a known chemical state while bubbles are passed through them. In addition, the main physical parameters, h_i, D_i, and D_s, are difficult to measure. Determinations of h_i and D_i commonly disagree with one another by factors of 2, and only two measurements of D_s seem to have been made. That of Sakata and Berg (1969) is probably in error because of their assumptions that bulk diffusion is always much faster than desorption and readsorption, even when times of several hours are in question, and that evaporation of their myristic acid could be neglected. We are left with Imahori's (1952) work, in which D_s for an adsorbed protein of relative molecular mass 70400 was found to be 1.1×10^{-10} m^2 sec^{-1}, or about twice its bulk diffusivity.

G. Stagnant Surfaces

In many practical applications, enough surfactant is present for rising bubbles and drops to have their surface motion almost completely stopped. Much of the experimental work gives nearly rigid-sphere values of drag coefficients for undistorted bubbles (Haberman and Morton, 1953, 1956; Peebles and Garber, 1953) and drops (Hu and Kintner, 1955; Licht and Narasimhamurty, 1955; Keith and Hixson, 1955; Klee and Treybal, 1956; Elzinga and Banchero, 1961). Because the transfer of heat and mass to bubbles and drops obeys different laws for free and rigid surfaces (Levich, 1962) it is important to know how small the velocity at the surface should be for the rigid-surface theory to apply.

At low Péclet numbers (much less than one) there is no problem: the diffusion equation is $\partial c/\partial t = D\nabla^2 c$ in either case, and for steady flow involving surfactants the theory is to be found in Section IV,D. But at high Péclet numbers the diffusion equation is no longer $4\partial c/\partial x = \partial^2 c/\partial y^2$ (see Eqs. 4.43 and 4.44), but (Levich, 1962)

$$9\partial c/\partial X = (1/Y)\partial^2 c/\partial Y^2, \tag{4.62}$$

for flow past a rigid sphere at low Reynolds numbers, for which

$$p_{r\theta} = 3U\eta_0 \sin\theta/2a \quad \text{on the surface,} \tag{4.63}$$

$$\psi \doteqdot \tfrac{3}{4}U(r-a)^2 \sin^2\theta \quad \text{near it,} \tag{4.64}$$

$$X = \theta - \tfrac{1}{2}\sin 2\theta, \quad \text{making} \quad dX/d\theta = 2\sin^2\theta, \tag{4.65}$$

$$\begin{aligned} Y &= 2\psi^{1/2}/3^{5/6}D_0^{1/3}U^{1/6}a^{2/3} \\ &\doteqdot (P_0/6)^{1/3}[(r-a)\sin\theta/a], \end{aligned} \tag{4.66}$$

where $P_0 = 2Ua/D_0 \gg 1$ as before. The solutions of Eq. (4.62) will normally vary significantly over ranges of Y of order unity, and so the boundary layer thickness will be of order $aP_0^{-1/3} \gg aP_0^{-1/2}$, the value for free surfaces. In the diffusion boundary layer, the fluid velocity is then $O(UP_0^{-1/3})$.

Like Eq. (4.43), Eq. (4.62) has some simple similarity solutions. We need the one which takes values $c = 0$ on $X = 0$ for $Y > 0$, $c \to 0$ as $Y \to \infty$ for $X > 0$, and $c = 1$ on $Y = 0$ for $X > 0$, namely

$$\begin{aligned} c &= \frac{\int_z^\infty \exp(-t^3)\,dt}{\int_0^\infty \exp(-t^3)\,dt} \\ &= \frac{1}{(\frac{1}{3})!}\int_z^\infty \exp(-t^3)\,dt = h(z), \text{ say,} \end{aligned} \tag{4.67}$$

where

$$z = Y/X^{1/3}. \tag{4.68}$$

With x, y, 2, and $\frac{1}{2}$ replacing X, Y, 3, $\frac{1}{3}$ throughout we recover the corresponding solution of Eq. (4.43).

Equations equivalent to (4.62)–(4.68) are the basis of Levich's (1962) diffusion theory for rigid spheres, and we must now follow Dukhin and Buikov (1965) by investigating what strength of surfactant solution makes bubbles (or drops) obey them. Equation (4.63) specifies the surface shear stress, and so

$$\begin{aligned}\Pi = R_g T h_0 c &= \Pi_0 + \tfrac{3}{2} U\eta_0 (1 - \cos\theta) \\ &= \Pi_0 + g(X), \text{ say,}\end{aligned} \tag{4.69}$$

on the surface of the bubble or drop, if Π_0 denotes the surface pressure at the front stagnation point. Equation (4.69) is one boundary condition to (4.62) in the region $0 \leq X \leq \pi$, $0 \leq Y < \infty$, the other being that $\Pi \to \Pi_\infty$, the fictitious surface pressure at a great distance defined by (4.30), as $Y \to \infty$ and as $X \to 0$ for $Y > 0$. The solution resembles Eq. (4.51), being

$$\Pi = \Pi_\infty + (\Pi_0 - \Pi_\infty) h\left(\frac{Y}{X^{1/3}}\right) + \int_0^X \frac{dg(t)}{dt} h\left(\frac{Y}{(X-t)^{1/3}}\right) dt, \tag{4.70}$$

where the functions g and h are defined in Eqs. (4.67) and (4.69). From Eq. (4.70) the value of $\partial\Pi/\partial r$ at the surface is

$$\begin{aligned}\frac{\partial\Pi}{\partial r} &= \left(\frac{P_0}{6}\right)^{1/3} \frac{\sin\theta}{a} \frac{\partial\Pi}{\partial Y} \\ &= \left(\frac{P_0}{6}\right)^{1/3} \frac{\sin\theta}{(\frac{1}{3})!\, a} \left(\frac{\Pi_\infty - \Pi_0}{X^{1/3}} - \int_0^X \frac{dg(t)}{dt} \frac{dt}{(X-t)^{1/3}}\right).\end{aligned} \tag{4.71}$$

Substitution of Π and $\partial\Pi/\partial r$ into Eq. (4.42),

$$a \frac{\partial}{\partial\theta}(\Pi u_\theta \sin\theta) = D_s \frac{\partial}{\partial\theta}\left(\sin\theta \frac{\partial\Pi}{\partial\theta}\right) + \frac{a^2 D_0 \sin\theta}{h_0} \frac{\partial\Pi}{\partial r},$$

allows us to calculate u_θ. For the theory to be consistent, u_θ must be much less than $UP_0^{-1/3}$, or

$$\Pi_0 \gg \max\left(\frac{D_s \eta_0 P_0^{1/3}}{a}, \frac{D_0 \eta_0 P_0^{2/3}}{h_0}\right). \tag{4.72}$$

To find Π_0, we integrate Eq. (4.42) right along the surface to obtain

$$\int_0^\pi \sin\theta \frac{\partial\Pi}{\partial r} d\theta = 0 = \int_0^\pi \frac{\partial\Pi}{\partial Y} dX, \tag{4.73}$$

and hence from (4.71)

$$\Pi_\infty - \Pi_0 = \frac{2U\eta_0}{\pi^{2/3}} \int_0^\pi \int_0^\theta \frac{\sin\theta' \sin^2\theta \, d\theta' \, d\theta}{(\theta - \theta' - \frac{1}{2}\sin 2\theta + \frac{1}{2}\sin 2\theta')^{1/3}},$$

$$\doteqdot 1.768 U\eta_0, \tag{4.74}$$

a result equivalent to that of Dukhin and Buikov (1965) to the degree of accuracy they used.

For a bubble or drop to behave like a rigid sphere as far as convective diffusion to its surface through the continuous phase is concerned, we see that $\Pi_\infty = R_g T h_0 c_\infty$ must be large enough for Π_0 given by (4.74) to obey the inequality (4.72). If Π_∞ is not this large, the surface velocity u_θ will be great enough to nullify equation (4.62) and make the boundary layer diffusion equation take a form like (2.27), i.e.

$$\partial c/\partial x = D_0 \, \partial^2 c/\partial \psi^2, \tag{4.75}$$

where ψ is the stream function and

$$x = \int_0^s \{m(s')\}^2 u_\theta(s') \, ds', \tag{4.76}$$

where s measures arc length from the front stagnation point ($a\theta$ for a sphere) and $m(s)$ is distance from the axis of symmetry to the bubble surface ($a \sin\theta$ for a sphere). Of course, this set of equations is difficult to work with if u_θ is one of the unknowns, and calculations can be found in the literature only for $u_\theta = v_0 \sin\theta$ (see Section IV,E), the form it takes for surfactants so dilute that the motion is hardly affected. Authors who assume $u_\theta = v_0 \sin\theta$ in other circumstances may be overoptimistic.

H. Insoluble Surfactants

So far, we have considered bubbles moving through solutions of surface-active material when their interfaces have reached a steady state, in which the total rate of diffusion to the more sparsely covered parts of their surfaces balances the rate of diffusion away from the more heavily covered parts. If the surfactant is very strong (i.e. the adsorption depths h_i are very large) it may happen that the latter process is very slow. (A characteristic time for it is evidently the smaller value of $h_i \delta_i / D_i$.) Then we may consider bubbles to be rising with a constant amount of surfactant adsorbed, and bulk diffusion can be neglected. If so, Eq. (4.42) simplifies to

$$a\Pi u_\theta = D_s \, \partial\Pi/\partial\theta, \tag{4.77}$$

if the bubble is spherical, and so Π increases by a factor e while θ increases by the amount au_θ/D_s, i.e.

$$\Pi = \Pi_0 \exp\left(\int_0^\theta \frac{au_\theta(\theta')}{D_s}\,d\theta'\right). \tag{4.78}$$

The surface shear stress $p_{r\theta}$ is $(1/a)(\partial\Pi/\partial\theta) = \Pi u_\theta/D_s$ if the surface excess is small enough, and $p_{r\theta}$ never exceeds $O(U\eta_0/a)$ at low Reynolds numbers. Therefore either Π or u_θ is small at each point around the bubble, and the bubble tends to reach a state in which u_θ is of order U over a part of the surface where Π is very small and may be neglected, whereas $u_\theta \ll U$ over the remainder. The transition region between the two parts has small angular extent at large Péclet numbers, but its details have not been elucidated.

Theories for the motion of such bubbles have been developed by several authors. They all consider the transition between a tangentially stress-free and rigid surface to occur suddenly at an angle $\theta = \theta^*$. Savic (1953) began this work both experimentally and theoretically, for bubbles. Griffith (1962) reported many observations on drops and suggested an approximate method for adapting Savic's theory to them. The theory for bubbles has since been improved by Davis and Acrivos (1966), who used a series expansion for the shear stress in the range where it was nonzero, and found the coefficients by minimizing the velocity in that range.

Davis and Acrivos found rather sudden transitions from free to rigid behavior of bubbles in a given liquid as their diameter decreased, with $n = -4$ in the notation of (4.58), whereas Levich's (1962) theory gives a more gradual change with $n = -\frac{3}{2}$. Experiments (Bond and Newton, 1928; Griffith, 1962) may give either type, and vary widely among themselves.

I. Slow Adsorption

Levich (1962) mentions the possibility that the transfer of adsorbed solute from regions of high to low surface excess through the bulk solution might be hindered by slowness of adsorption or desorption, i.e. that there might be an "activation energy" barrier between the dissolved and adsorbed states of a surfactant. His equation (74.18) implies that $n = -1$ in our Eq. (4.58) for a surfactant whose energy barrier controls its adsorption. In this review such barriers have been ignored. We take the view that no convincing experimental evidence for their existence has yet been put forward, except possibly for systems with several interacting surface-active components.

Many workers who have measured the surface tension of freshly formed liquid surfaces disagree, but Defay and Hommelen (1958a) found no measurements on surfactant solutions in which proper allowance was made for the effect of the fluid motion on the surfactant distribution. It is, of course, impossible to alter the area of a fluid surface without setting the fluid in motion.

Hansen *et al.* (1958), Defay and Pétré (1962), and Hansen (1964) have made a start on determining short-term hydrodynamical effects in vibrating-jet measurements of surface tension. It seems that surface ages can be overestimated by a millisecond or so, and surface tensions can be overestimated by a few percent in the first few milliseconds. This casts some doubt on the suggestion of Tsonopoulos *et al.* (1971) that Defay and Hommelen's (1958b) experimental results imply an adsorption barrier for hexanol. For surface ages of several seconds the hydrodynamics may no longer be a problem, but evaporation of the solute may simulate an adsorption barrier with higher alcohols such as decanol (Defay and Hommelen, 1959b).

The above comments apply to systems in which chemical reactions between the components can be ignored. When several reacting surface-active substances are present, it may happen that an element of fluid in the bulk solution contains an equilibrium mixture, but the mixture is no longer in chemical equilibrium if brought to the surface when each component adsorbs according to its own value of $h = \Gamma/c$. If so, the approach to surface equilibrium may be delayed by the slowness of the reactions restoring equilibrium among the various chemical components. Hansen and Wallace (1959) found some evidence for this in organic acids, which are known to exist in ionized, unionized, and associated forms. Defay and Hommelen (1959b) found that a slow reaction occurred for the dibasic azelaic acid $HO_2C(CH_2)_7CO_2H$.

Reaction times of the order of milliseconds are the most important for bubbles, because a bubble small enough to be appreciably affected by surfactants rises through its own diameter in a few milliseconds in water. Unfortunately, there is no very good method of measuring surface tensions which vary as rapidly as this (Defay and Hommelen, 1958a, 1959a; Defay and Petre, 1962; Wegener and Parlange, 1964; reviewed by Defay *et al.*, 1966). Another complication is that dissolved ions, especially polyvalent ones, may modify the molecular structure of the water in their neighborhood enough to affect the behavior of bubbles. Zieminski and Whittemore (1971) discussed their experiments on bubble coalescence from this point of view, but admit that it is controversial.

ACKNOWLEDGMENTS

This article was written at Imperial College, London, during a period of refresher leave from the Victoria University of Wellington, New Zealand, and I am most grateful to both institutions. I also wish to thank Prof. G. Astarita, Prof. G. K. Batchelor, Dr. T.

Blake, Prof. A. V. Bradshaw, Dr. R. Collins, Dr. A. R. H. Cornish, Dr. G. J. Jameson, Dr. J. A. K. Kitchener, Dr. H. Sawitowski, Prof. L. E. Scriven, and especially Dr. D. W. Moore for discussions, Miss J. Pindelska, Librarian, Mathematics Department, Imperial College, for help in obtaining and copying references, and the Cambridge University Press for permitting Fig. 7, part of Table 1, and Table 3 to be copied from papers in the *Journal of Fluid Mechanics*.

REFERENCES

ABRAMOWITZ, M., and STEGUN, I. A. (1965). "Handbook of Mathematical Functions." Dover, New York.

ACKERET, J. (1952). Über exakte Lösungen der Stokes-Navier-Gleichungen inkompressibler Flüssigkeiten bei veränderten Grenzbedingungen. *Z. Angew. Math. Phys.* **3**, 259–271.

ADAM, N. K. (1968). "The Physics and Chemistry of Surfaces." Dover, New York.

ANGELINO, H. (1966). Hydrodynamique des grosses bulles dans les liquides visqueux. *Chem. Eng. Sci.* **21**, 541–550.

ASTARITA, G. (1970). Personal communication.

BAKKER, C. A. P., VAN BUYTENEN, P. M., and BEEK, W. J. (1966). Interfacial phenomena and mass transfer. *Chem. Eng. Sci.* **21**, 1039–1046.

BATCHELOR, G. K. (1956). On steady laminar flow with closed streamlines at large Reynolds number. *J. Fluid Mech.* **1**, 177–190.

BIRD, W. B., STEWART, W. E., and LIGHTFOOT, E. N. (1960). "Transport Phenomena." Wiley, New York.

BIRKHOFF, G., and ZARANTONELLO, E. H. (1957). "Jets, Wakes, and Cavities." Academic Press, New York.

BOND, W. N., and NEWTON, D. A. (1928). Bubbles, drops and Stokes' law. (Paper 2.) *Phil. Mag.* [7] **5**, 794–800.

BRIGNELL, A. S. (1970). Ph.D. Thesis, Imperial College, London University.

BRODKEY, R. S. (1967). "The Phenomena of Fluid Motions." Addison-Wesley, Reading, Massachusetts.

BRYN, T. (1933). Steiggeschwindigkeit von Luftblasen in Flüssigkeiten. *Forsch. Geb. Ingenieurw.* **4**, 27–30.

CHAN, B. K. C., and PRINCE, R. G. H. (1965). Distillation studies—viscous drag on a gas bubble rising in a liquid. *AIChEJ* **11**, 268–273.

CHAO, B. T. (1962). Motion of spherical gas bubbles in a viscous fluid at large Reynolds numbers. *Phys. Fluids* **5**, 69–79.

CHESTER, W., and BREACH, D. R. (1969). On the flow past a sphere at low Reynolds number. *J. Fluid Mech.* **37**, 751–760.

COLE, R. (1967). Motion of vapor bubbles in saturated liquids. *AIChEJ* **13**, 403–404.

COLLINS, R. (1966). A second approximation for the velocity of a large bubble rising in an infinite liquid. *J. Fluid Mech.* **25**, 469–480.

DAVENPORT, W. G., BRADSHAW, A. V., and RICHARDSON, F. D. (1967a). Behaviour of spherical cap bubbles in liquid metals. *J. Iron Steel Inst., London* **205**, 1034–1042.

DAVENPORT, W. G., RICHARDSON, F. D., and BRADSHAW, A. V. (1967b). Spherical cap bubbles in low density liquids. *Chem. Eng. Sci.* **22**, 1221–1235.

DAVIES, J. T., and RIDEAL, E. K. (1963). "Interfacial Phenomena," 2nd. ed. Academic Press, New York.

DAVIES, R. M., and TAYLOR, G. I. (1950). The mechanics of large bubbles rising through extended liquids and through liquids in tubes. *Proc. Roy. Soc., Ser. A* **200**, 375–390.

DAVIS, R. E., and ACRIVOS, A. (1966). The influence of surfactants on the creeping motion of small bubbles. *Chem. Eng. Sci.* **21**, 681–685.

DEFAY, R., and HOMMELEN, J. (1958a). Bibliographie critique sur les methodes de mesure des tensions superficielles dynamiques. *Ind. Chim. Belge* **23**, 597–614.

DEFAY, R., and HOMMELEN, J. (1958b). Measurement of dynamic surface tensions of aqueous solutions by the oscillating jet method. *J. Colloid Sci.* **13**, 553–564.

DEFAY, R., and HOMMELEN, J. (1959a). Measurement of dynamic surface tensions of aqueous solutions by the falling meniscus method. *J. Colloid Sci.*, **14**, 401–410.

DEFAY, R., and HOMMELEN, J. (1959b). The importance of diffusion in the adsorption process of some alcohols and acids in dilute aqueous solutions. *J. Colloid Sci.* **14**, 411–418.

DEFAY, R., and PÉTRÉ, G. (1962). Correcting surface tension data obtained by the oscillating jet method. *J. Colloid Sci.* **17**, 565–569.

DEFAY, R., PRIGOGINE, I., BELLEMANS, A., and EVERETT, D. H. (1966). "Surface Tension and Adsorption." Longmans, Green, New York.

DERJAGUIN, B. V., and DUKHIN, S. S. (1961). Theory of flotation of small and medium-size particles. *Inst. Mining Met., Trans.* **70**, 221–246.

DERYAGIN, B. V. (= DERJAGUIN, B. V.), DUKHIN, S. S., and LISICHENKO, V. A. (1959). The kinetics of the attachment of mineral particles to bubbles during flotation. I. The electric field of a moving bubble. *Russ. J. Phys. Chem.* **33**, 389–393.

DERYAGIN, B. V., DUKHIN, S. S., and LISICHENKO, V. A. (1960). The kinetics of the attachment of mineral particles to bubbles during flotation. II. The electric field of a moving bubble when the surface activity of the ionogenic substance is high. *Russ. J. Phys. Chem.* **34**, 248–251.

DUKHIN, S. S. (1966). An electrical diffusion theory of the Dorn effect and a discussion of the possibility of measuring zeta potentials. *Res. Surface Forces, Proc. Conf., 2nd,* 1962 Vol. 2, pp. 54–74.

DUKHIN, S. S., and BUIKOV, M. V. (1965). Theory of the dynamic adsorption layer of moving spherical particles. II. Theory of the dynamic adsorption layer of a bubble (drop) at a Reynolds number $\mathrm{Re} \leqslant 1$ in the presence of strong retardation of the surface. *Russ. J. Phys. Chem.* **39**, 482–485.

DUKHIN, S. S., and DERYAGIN, B. V. (1961). Kinetics of the attachment of mineral particles to bubbles during flotation. V. Motion of the bubble surface strongly retarded by surface-active substances. Distribution of the surface-active material and electric field of the bubble. *Russ. J. Phys. Chem.* **35**, 715–717.

DURHAM, K., ed. (1961). "Surface Activity and Detergency." Macmillan, New York.

EL SAWI, M. (1970). Ph.D. Thesis, Imperial College, London University.

ELZINGA, E. R., and BANCHERO, J. T. (1961). Some observations on the mechanics of drops in liquid-liquid systems. *AIChEJ* **7**, 394–399.

FRAENKEL, L. E. (1970). On steady vortex rings of small cross-section in an ideal fluid. *Proc. Roy. Soc., Ser. A* **316**, 29–62.

GAL-OR, B., KLINZING, G. E., and TAVLARIDES, L. L. (1969). Bubble and drop phenomena. *Ind. Eng. Chem.* **61**, 2, 21–34.

GARABEDIAN, P. R. (1957). On steady-state bubbles generated by Taylor instability. *Proc. Roy. Soc., Ser. A* **241**, 423–431.

GILL, W. N., COLE, R., ESTRIN, J., and NUNGE, R. J. (1969). Fluid dynamics. *Ind. Eng. Chem.* **61**, 1, 41–75.

GILL, W. N., COLE, R., DAVIS, E. J., ESTRIN, J., NUNGE, R. J., and LITTMAN, H. (1970). Fluid dynamics. *Ind. Eng. Chem.* **62**, 4, 49–79.

GLASSTONE, S. (1953). "The Elements of Physical Chemistry." Macmillan, New York.

GOLDSTEIN, S., ed. (1938). "Modern Developments in Fluid Dynamics," 2 vols. Cambridge Univ. Press, London and New York.

GOMEZPLATA, A., and REGAN, T. M. (1968). Mass transfer. *Ind. Eng. Chem.* **60**, 12, 53–62.

GOUSE, S. W., JR. (1966). "An Index to the Two-phase Gas-liquid Flow Literature." MIT Press, Cambridge, Massachusetts.

GRIFFITH, R. M. (1962). The effect of surfactants on the terminal velocity of drops and bubbles. *Chem. Eng. Sci.* **17**, 1057–1070.

GUGGENHEIM, E. A. (1949). "Thermodynamics." North-Holland Publ., Amsterdam.

GUTHRIE, R. I. L., and BRADSHAW, A. V. (1969). The stability of gas envelopes trailed behind large spherical cap bubbles rising through viscous liquids. *Chem. Eng. Sci.* **24**, 913–917.

HABERMAN, W. L., and MORTON, R. K. (1953). An experimental investigation of the drag and shape of air bubbles rising in various liquids. *David Taylor Model Basin Rep.* No. 802.

HABERMAN, W. L., and MORTON, R. K. (1956). An experimental study of bubbles moving in liquids. *Trans. Amer. Soc. Civil Eng.* **121**, 227–252.

HADAMARD, J. (1911). Mouvement permanent lent d'une sphère liquide et visqueuse dans une liquide visqueux. *C. R. Acad. Sci.* **152**, 1735–1738.

HAMMERSLEY, J. H. (1961). On the statistical loss of long-period comets from the solar system. II. *Proc. Symp. Math. Stat. Probability*, 4*th*, 1960, Vol. 3, pp. 17–78.

HANSEN, R. S. (1964). The calculation of surface age in vibrating jet measurements. *J. Phys. Chem.* **68**, 2012–2014.

HANSEN, R. S., and WALLACE, T. C. (1959). The kinetics of adsorption of organic acids at the water-air interface. *J. Phys. Chem.* **63**, 1085–1091.

HANSEN, R. S., PURCHASE, M. E., WALLACE, T. C., and WOODY, R. W. (1958). Extension of the vibrating jet method for surface tension measurement to jets of nonuniform velocity profiles. *J. Phys. Chem.* **62**, 210–214.

HAPPEL, J., and BRENNER, H. (1965). "Low Reynolds Number Hydrodynamics." Prentice-Hall, Englewood Cliffs, New Jersey.

HARPER, J. F. (1970). Viscous drag in steady potential flow past a bubble. *Chem. Eng. Sci.* **25**, 342–343.

HARPER, J. F. (1971). Errata to "Viscous drag in steady potential flow past a bubble." *Chem. Eng. Sci.* **26**, 501.

HARPER, J. F. (1972). On spherical bubbles rising steadily in dilute surfactant solutions. (to be published).

HARPER, J. F., and MOORE, D. W. (1968). The motion of a spherical liquid drop at high Reynolds number. *J. Fluid Mech.* **32**, 367–391.

HARPER, J. F., MOORE, D. W., and PEARSON, J. R. A. (1967). The effect of the variation of surface tension with temperature on the motion of bubbles and drops. *J. Fluid Mech.* **27**, 361–366.

HARTUNIAN, R. A., and SEARS, W. R. (1957). On the instability of small gas bubbles moving uniformly in various liquids. *J. Fluid Mech.* **3**, 27–47.

HILL, M. J. M. (1894). On a spherical vortex. *Phil. Trans. Roy. Soc. London, Ser. A* **185**, 213–245.

HU, S., and KINTNER, R. C. (1955). The fall of single liquid drops through water. *AIChEJ.* **1**, 42–48.

IMAHORI, K. (1952). The structure of surface-denatured protein. III. Determination of the shape of surface-denatured horse serum albumin. *Bull. Chem. Soc. Jap.* **25**, 13–16.

JONES, D. R. M. (1965). Ph.D. Thesis, University of Cambridge.

Keith, F. W., and Hixson, A. N. (1955). Liquid-liquid extraction spray columns. *Ind. Eng. Chem.* **47**, 258–267.

Kendall, D. G. (1961). Some problems in the theory of comets. II. *Proc. Symp. Math. Stat. Probability*, 4*th*, 1960 Vol. 3, pp. 121–147.

Kenning, D. B. R. (1969). The effect of surface energy variations on the motion of bubbles and drops. *Chem. Eng. Sci.* **24**, 1385–1386.

Klee, A. J., and Treybal, R. E. (1956). Rate of rise or fall of liquid drops. *AIChEJ.* **2**, 444–447.

Kochina, N. N. (1969). Some solutions of the inhomogeneous heat equation. (In Russian.) *Dokl. Akad. Nauk SSSR* **186**, 54–57; see *Sov. Phys.—Dokl.* **14**, 419–421.

Kojima, E., Akehata, T., and Shirai, T. (1968). Rising velocity and shape of single air bubbles in highly viscous liquids. *J. Chem. Eng. Jap.* **1**, 45–50.

Lagerstrom, P. A., and Cole, J. D. (1955). Examples illustrating expansion procedures for the Navier-Stokes equations. *J. Rat. Mech. Anal.* **4**, 817–882.

Lamb, H. (1932). "Hydrodynamics," 6th ed. Cambridge Univ. Press, London and New York.

Landau, L. D., and Lifshitz, E. M. (1959). "Fluid Mechanics." Pergamon, Oxford.

Lane, W. R., and Green, H. L. (1956). The mechanics of drops and bubbles. *In* "Surveys in Mechanics" (G. K. Batchelor and R. M. Davies, eds.), pp. 162–215. Cambridge Univ. Press, London and New York.

Levich, V. G. (1949). Motion of gaseous bubbles with high Reynolds numbers. (In Russian.) *Zhn. Eksp. Teor. Fiz.* **19**, 18–24.

Levich, V. G. (1962). "Physicochemical Hydrodynamics." Prentice-Hall, Englewood Cliffs, New Jersey.

Licht, W., and Narasimhamurty, G. S. R. (1955). Rate of fall of single liquid droplets. *AIChEJ.* **1**, 366–373.

Lighthill, M. J. (1958). On displacement thickness. *J. Fluid Mech.* **4**, 383–392.

Lochiel, A. C. (1965). The influence of surfactants on mass transfer around spheres. *Can. J. Chem. Eng.* **43**, 40–44.

McGlashan, M. L. (1968). "Physico-Chemical Quantities and Units." Roy. Inst. Chem., London.

Malenkov, E. G. (1968). Motion of large gas bubbles rising into fluids (in Russian). *Zh. Prikl. Mekh. Tekh. Fiz.* **6**, 130–133; to be translated in *J. Appl. Mech. Tech. Phys.* (*USSR*).

Maxworthy, T. (1967). A note on the existence of wakes behind large, rising bubbles. *J. Fluid Mech.* **27**, 367–368 and 2 plates.

Melrose, J. C. (1968). Thermodynamic aspects of capillarity. *Ind. Eng. Chem.* **60**, 3, 53–70.

Mendelson, H. D. (1967). The prediction of bubble terminal velocities from wave theory. *AIChEJ.* **13**, 250–253.

Moore, D. W. (1959). The rise of a gas bubble in a viscous liquid. *J. Fluid Mech.* **6**, 113–130.

Moore, D. W. (1963). The boundary layer on a spherical gas bubble. *J. Fluid Mech.* **16**, 161–176.

Moore, D. W. (1965). The velocity of rise of distorted gas bubbles in a liquid of small viscosity. *J. Fluid Mech.* **23**, 749–766.

Newman, J. (1967). Retardation of falling drops. *Chem. Eng. Sci.* **22**, 83–85.

Pan, Y. F., and Acrivos, A. (1968). Shape of a drop or bubble at low Reynolds number. *Ind. Eng. Chem., Fundam.* **7**, 227–232.

Parlange, J. Y. (1969). Spherical bubbles with laminar wakes. *J. Fluid Mech.* **37**, 257–263.

PARSONS, R. (1954). Equilibrium properties of electrified interphases. *In* "Modern Aspects of Electrochemistry" (J. O'M. Bockris, ed.), Vol. 1, pp. 103–179. Butterworth, London.

PEARSON, J. R. A. (1958). On convection cells induced by surface tension. *J. Fluid Mech.* **4**, 489–500.

PEEBLES, F. N., and GARBER, H. J. (1953). Studies on the motion of gas bubbles in liquids. *Chem. Eng. Progr.* **49**, 88–97.

PERRY, R. H., CHILTON, C. H., and KIRKPATRICK, S. D., eds. (1963). "Chemical Engineers Handbook," 4th ed. McGraw-Hill, New York.

PLESSET, M. S. (1964). Bubble dynamics. *In* "Cavitation in Real Liquids" (R. Davies, ed.), pp. 1–18. Elsevier, Amsterdam.

PRANDTL, L. (1905). Über Flüssigkeitsbewegung bei sehr kleiner Reibung. *Verh. Int. Math.-Kongr., 3rd*, 1904 pp. 484–491.

PRANDTL, L., and TIETJENS, O. G. (1957). "Applied Hydro- and Aerodynamics." Dover, New York.

PROUDMAN, I., and PEARSON, J. R. A. (1957). Expansions at small Reynolds numbers for the flow past a sphere and a circular cylinder. *J. Fluid Mech.* **2**, 237–262.

PRUPPACHER, H. R., LE CLAIR, B. P., and HAMIELEC, A. E. (1970). Some relations between drag and flow pattern of viscous flow past a sphere and a cylinder at low and intermediate Reynolds numbers. *J. Fluid Mech.* **44**, 781–790.

REDFIELD, J. A., and HOUGHTON, G. (1965). Mass transfer and drag coefficients for single bubbles at Reynolds numbers of 0.2–5000. *Chem. Eng. Sci.* **20**, 131–139.

REGAN, T. M., and GOMEZPLATA, A. (1970). Mass transfer. *Ind. Eng. Chem.* **62**, 2, 41–53.

RIPPIN, D. W. T., and DAVIDSON, J. F. (1967). Free streamline theory for a large gas bubble in a liquid. *Chem. Eng. Sci.* **22**, 217–227.

ROSENHEAD, L., ed. (1963). "Laminar Boundary Layers." Oxford Univ. Press, London and New York.

RUBIN, E., and JORNE, J. (1969). Foam separation of solutions containing two ionic surface-active agents. *Ind. Eng. Chem., Fundam.* **8**, 474–482.

RUCKENSTEIN, E. (1967). Mass transfer between a single drop and a continuous phase. *Int. J. Heat Mass Transfer* **10**, 1785–1792.

RYBCZYNSKI, W. (1911). On the translatory motion of a fluid sphere in a viscous medium (in German). *Bull. Int. Acad. Pol. Sci. Lett. Cl. Sci. Math. Natur., Ser. A* pp. 40–46.

SAFFMAN, P. G. (1956). On the rise of small air bubbles in water. *J. Fluid Mech.* **1**, 249–275.

SAKATA, E. K., and BERG, J. C. (1969). Surface diffusion in monolayers. *Ind. Eng. Chem., Fundam.* **8**, 570–575.

SAVIC, P. (1953). Circulation and distortion of liquid drops falling through a viscous medium. *Nat. Res. Counc. Can., Div. Mech. Eng. Rep.* **MT-22**.

SCHMIDT, E. (1933). Über die Bewegung von Flüssigkeits-Gas-Gemischen. *VDI* (*Ver. Deut. Ing.*) *Z.* **77**, 1159–1160.

SCRIVEN, L. E. (1960). Dynamics of a fluid interface. *Chem. Eng. Sci.* **12**, 98–108.

SCRIVEN, L. E., and STERNLING, C. V. (1960). The Marangoni effects. *Nature* (*London*) **187**, 186–188.

SCRIVEN, L. E., and STERNLING, C. V. (1964). On cellular convection driven by surface-tension gradients: Effects of mean surface tension and surface viscosity. *J. Fluid Mech.* **19**, 321–340.

SHEDLOVSKY, L. (1968). Interface properties of detergent films. *Ind. Eng. Chem.* **60**, 10, 47–52.

SHINODA, K., and NAKANISHI, J. (1963). Selective adsorption studies by radio tracer techniques. IV. Selective adsorptivity of alcohol against surface active agent at the air-solution interface. *J. Phys. Chem.* **67**, 2547–2549.

Shoemaker, P. D., and Marc de Chazal, L. E. (1969). Dimpled and skirted liquid drops moving through viscous liquid media. *Chem. Eng. Sci.* **24**, 795–797.

Slaughter, I., and Wraith, A. E. (1968). The wake of a large gas bubble. *Chem. Eng. Sci.* **23**, 932 and plate.

Stahel, E. P., and Ferrell, J. K. (1968). Heat transfer. *Ind. Eng. Chem.* **60**, 1, 75–83.

Stewartson, K. (1968). On an integral equation. *Mathematika* **15**, 22–29.

Sumner, B. S., and Moore, F. K. (1969). Comments on the nature of the outside boundary layer on a liquid sphere in a steady, uniform stream. *NASA Contract. Rep.* **NASACR-1362.**

Taunton, J. W., and Lightfoot, E. N. (1969). Transient effects in mass transfer to circulating drops. *Int. J. Heat Mass Transfer* **12**, 1718–1722.

Tavlarides, L. L., Coulaloglou, C. A., Zeitlin, M. A., Klinzing, G. E., and Gal-Or, B. (1970). Bubble and drop phenomena. *Ind. Eng. Chem.* **62**, 11, 6–27.

Taylor, T. D., and Acrivos, A. (1964). On the deformation and drag of a falling viscous drop at low Reynolds number. *J. Fluid Mech.* **18**, 466–476.

Thomson, J. J., and Newall, H. F. (1885). On the formation of vortex rings by drops falling into liquids, and some allied phenomena. *Proc. Roy. Soc.* **39**, 417–437.

Thorsen, G., Stordalen, R. M., and Terjesen, S. G. (1968). On the terminal velocity of circulating and oscillating liquid drops. *Chem. Eng. Sci.* **23**, 413–426.

Tsonopoulos, C., Newman, J., and Prausnitz, J. M. (1971). Rapid aging and dynamic surface tension of dilute aqueous solutions. *Chem. Eng. Sci.* **26**, 817–827.

Tyroler, G., Hamielec, A. E., Johnson, A. I., and Leclair, B. P. (1971). Mass transfer with fast chemical reaction in drops. *Can. J. Chem. Eng.* **49**, 56–61.

Walbridge, N. L., and Woodward, L. A. (1970). Phase velocity of surface capillary-gravity waves. *Phys. Fluids* **13**, 2461–2465.

Wasserman, M. L., and Slattery, J. C. (1969). Creeping flow past a fluid globule when a trace of surfactant is present. *AIChEJ.* **15**, 533–547.

Weatherburn, C. E. (1927). "Differential Geometry of Three Dimensions," Vol. I. Cambridge Univ. Press, London and New York.

Wegener, P. P., and Parlange, J.-Y. (1964). Surface tension of liquids from water-bell experiments. *Z. Phys. Chem. (Frankfurt am Main)* **43**, 245–259.

Winnikow, S., and Chao, B. T. (1966). Droplet motion in purified systems. *Phys. Fluids* **9**, 50–61.

Young, N. O., Goldstein, J. S., and Block, M. J. (1959). The motion of bubbles in a vertical temperature gradient. *J. Fluid Mech.* **6**, 350–356.

Zieminski, S. A., and Whittemore, R. C. (1971). Behavior of gas bubbles in aqueous electrolyte solutions. *Chem. Eng. Sci.* **26**, 509–520.

Shock Waves, Jump Relations, and Structure*

P. GERMAIN

University of Paris (VI), Paris, France

* This paper is based on the text of lectures given by the author at "The International School of Nonlinear Mathematics and Physics" held in Munich.

Introduction

A theory is a mathematical scheme of a physical phenomenon. Here, we will be concerned with phenomena which can be described by a nonlinear system of partial differential equations. But in fact a single physical phenomenon may be described by different theories, according to the refinement one wants to put into the mathematical description. The consistency of the hierarchy of theories implies that the results given by one of these mathematical schemes may be recovered as a limiting case of those given by a more refined one. From a physical point of view it is desirable to take into account, in the formulation of the mathematical theory, all the aspects of the physical process, but with a simpler representation one can often go further in the mathematical resolution. Thus, it is always interesting to consider different theories of increasing complexity in order to describe a single physical phenomenon.

Most of the nonlinear problems related to wave phenomena are described in the *simplified theory* by a mathematical system (SP) which does not always admit a *classical solution* (cs) (for instance, continuously differentiable) even if, from a physical point of view, they are "well posed." The solution which has to be considered may be discontinuous. It belongs to a class (ss) of solutions of the mathematical problem which may be called the *shock solutions* of the problem, because the most interesting manifolds along which the solution is discontinuous are called *shock waves*. This class (ss) is a *strict subset* of the class of *weak solutions* (ws), and, obviously, the set of the classical solutions (cs) is itself a strict subset of the class of shock solutions. One of the most stimulating tasks of mathematical physics is to show that a nonlinear problem admits one and only one solution which belongs to this class and that this solution depends continuously on the data.

Now, quite often, for the same process, described in a *more refined scheme*, there exists a classical solution for the new and more complicated mathematical problem (DP). From the physical point of view, at least in many cases, the simplified version (SP) is obtained when all dissipation effects are neglected in this problem (DP). Thus, a shock solution of (SP) must be, in some sense, the limit of a classical solution of a related (DP). When dissipation effects tend towards zero, obviously the convergence cannot be uniform, in the neighborhood of a shock wave; in other words, (DP) is a singular perturbation of the related (SP). Then, such a shock appears as the limiting result of a narrow region in which some physical quantities vary very rapidly. The study of the behavior of the solution of (DP) in such a region for very small dissipation effects is what is called the study of the *structure of the shock*. The knowledge of the structure puts some light on the physical nature of a shock wave: *it is the region where nonlinear effects of convection which steepen the gradients of physical characteristics of the problem are balanced by the diffusion effects due to vanishingly small dissipation.*

In the present paper, we will be concerned only with the local study of shock waves. Section I deals with a simple mathematical model related to a single conservation law. Section II is devoted to the "shock relations" for a physical phenomenon governed by conservation laws and the classification of shocks. Examples will be given in some detail, in particular taken from classical gas dynamics and magnetofluid dynamics. Problems related to the structure of a shock are considered in Section III, where some examples are given in order to show that quite different behaviors may be found. Finally, in Section IV are given asymptotic expansions, which describe completely the behavior of the solution in the neighborhood of a shock. For simplicity, the analysis is restricted to the case of classical gas dynamics. It is shown that the classical results of Section II and III are in fact "the first terms" of the expansions given in this chapter. One may say that in Section III the consistency of different theories describing the same physical phenomenon is investigated—as far as shocks are concerned—and that Section IV shows how the classical theory of shocks is in fact for a given physical theory a coherent mathematical first-order approximation. Briefly speaking, this paper is an attempt to support the mathematical and physical validity of the concept of shock waves.

I. Shock Solution for a Mathematical Model Related to a Single Conservation Law

The purpose of this section is to introduce the basic concepts and ideas which arise in the theory of shocks. For simplicity they will be given for a very simple mathematical model. Most of the results we need are classical and may be found in the literature, so that proofs will generally be omitted.

A. The Mathematical Model—Classical Solutions (cs)

The nonlinear partial differential equation we want to study describes a plane wave phenomenon. We will take

$$\frac{\partial u}{\partial t} + u\frac{\partial u}{\partial x} = 0, \tag{1.1}$$

or, more generally,

$$\frac{\partial u}{\partial t}+\frac{\partial}{\partial x}f(u)=0, \tag{1.2}$$

where $f(u)$ is a twice continuously differentiable function strictly convex—$f''(u)>0$—of the single scalar variable u, which is supposed to denote a physical quantity. Equation (1.1) is the special case of (1.2) when $f(u)=u^2/2$. The independent variables t and x are respectively the time and space coordinates; thus for instance in (1.1) u is a velocity. We look for the solution of a *Cauchy problem*:

Let a function $u_0(x)$ be given; find, for $t\geq 0$, the unknown function $u(x, t)$ such that $u(x, 0)=u_0(x)$.

The classical general solution of (1.2) is defined implicitly by

$$x-f'(u)t=\varphi(u), \tag{1.3}$$

where $\varphi(u)$ is an arbitrary function. In the three-dimensional half-space $(x, t, u;\ t\geq 0)$ it represents a conoid, generated by a set of straight lines, $u=$ const., parallel to the x, t plane. The classical solution of the Cauchy problem is then defined by the conoid, generated by the one-parameter family of straight lines

$$u=u_0(\xi); \qquad x-f'(u_0(\xi))t=\xi. \tag{1.4}$$

If the unknown is the *surface* of the (x, t, u) space, which satisfies (1.2) and the given conditions for $t=0$, then (1.4) gives the existence and the uniqueness of this solution.

But, as we are interested in a plane wave phenomenon, such an interpretation is not convenient if u is a physical quantity. For, in a convenient physical interpretation, u must be a single-valued function of x and t. This implies that for every $t>0$, the mapping $\xi\rightarrow f'(u_0(\xi))t+\xi$ must be one to one. If $u_0(x)$ is differentiable, this condition requires $u_0'(x)\geq 0$. Conversely, if $u_0'(x)\geq 0$, the classical solution fulfills all the requirements.

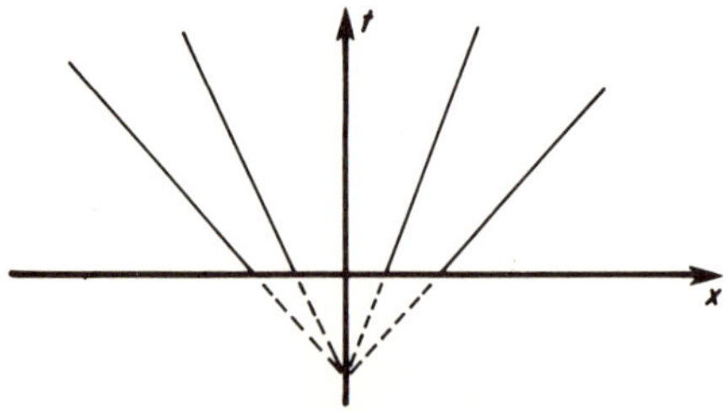

FIG. 1. Equation (1.1): $u_0=x\rightarrow u=x/(1+t)$.

For instance, for Eq. (1.1), if $u_0(x) = x$, the solution is $u = x(1+t)^{-1}$. The classical solution is also the correct one for any nondecreasing $u_0(x)$; for there exists at most an enumerable set of points of discontinuity and if x_0 is such a point one gets

$$u(x, t) = g\left(\frac{x - x_0}{t}\right)$$

for

$$x_0 + f'[u_0(x_0 - 0)]t \leq x \leq x_0 + f'[u_0(x_0 + 0)]t,$$

where $u = g(\tau)$ is the reciprocal increasing function of $\tau = f'(u)$; see the example of Fig. 2. Note that, along a line of constant u, $u_x(x, t)$ is a

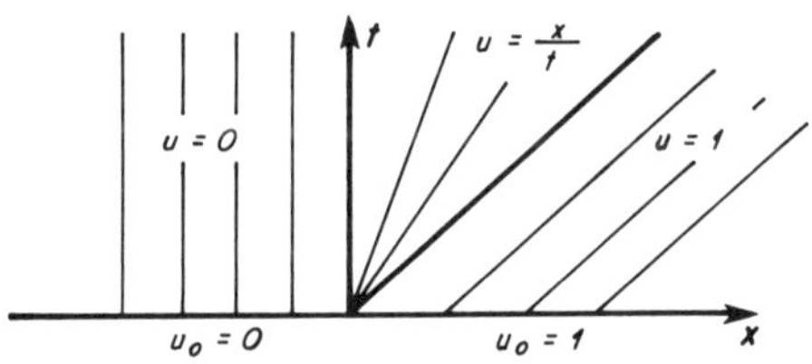

FIG. 2. Equation (1.1): $u_0 = 0$ at $x < 0$. $u_0 = 1$ at $x > 0$.

decreasing function of t. Now if $u_0(x)$ is not a nondecreasing function, the classical solution is not a single valued function and thus cannot be the correct physical solution. Take Eq. (1.1) for instance, the characteristics have a real envelope (L):

$$t = -\frac{1}{u_0'(\xi)}, \qquad x = \xi - \frac{u_0(\xi)}{u_0'(\xi)},$$

which is called the "limiting line" of the solution, (see, for example, Fig. 3). If ξ_m is the value which gives the minimum (negative) value of

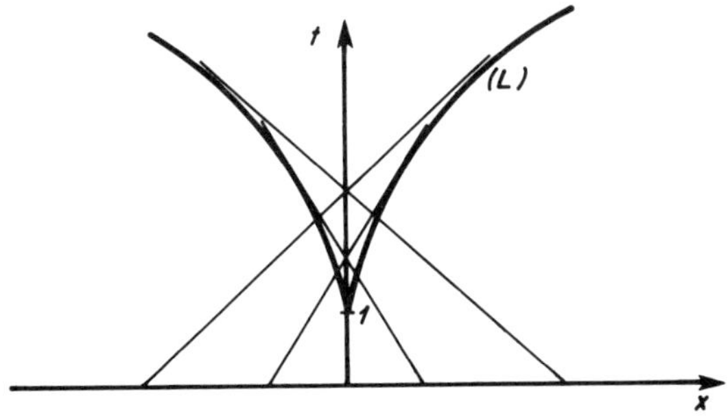

FIG. 3. Equation (1.1): $u_0 = -\text{th}\, x$; $t = \text{ch}^2 \xi$ and $x = \xi - \text{sh}\, \xi\, \text{ch}\, \xi$ define the limiting line (L).

$u'(x)$, then the solution is single valued for $0 \leq t \leq -[u_0'(\xi_m)]^{-1} = t_m$. For $t \leq t_m$ the absolute value of the slope of the graph $u(x, t)$, for t fixed, is an increasing function of t along a characteristic and the solution ceases to be valid for $t = t_m$ when one of the slopes becomes infinite. For $t > t_m$ there exist points in the x, t half-plane ($t > 0$) through which run at least two characteristics and the classical solution becomes multivalued. Then the nonlinear convection has an *attenuation* effect when $u_0'(x)$ is positive, and a *steepening* effect when $u_0'(x)$ is negative. In the latter case the classical solution cannot be valid for every t.

B. Weak Solutions (ws)

1. *Definition*

The field (cs) is too narrow to describe physical phenomena governed by Eq. (1.2). Therefore, to widen this field, we can consider the *weak solutions* of (1.2). For a numerical summable function $u(x, t)$, it is possible to compute the left member of (1.2), when considering u and $f(u)$ as distributions. Then $u(x, t)$ is a weak solution of the given Cauchy problem if $u = 0$ for $t < 0$ and if

$$\frac{\partial u}{\partial t} + \frac{\partial}{\partial x} f(u) = u_0(x)\,\delta(t).$$

That means that (Hopf, 1950; Lax, 1957; Germain, 1960a) if $\alpha(x, t)$ is a continuously differentiable function which is zero outside a compact set of the x, t plane:

$$\int_0^\infty dt \int_{-\infty}^\infty \left\{\frac{\partial \alpha}{\partial t} u + \frac{\partial \alpha}{\partial x} f(u)\right\} dx + \int_{-\infty}^\infty \alpha(x, 0) u_0(x)\, dx = 0. \tag{1.5}$$

One can show that another equivalent definition is as follows: $u(x, t)$ is a weak solution if $u(x, 0) = u_0(x)$ almost everywhere, and if for every closed contour (c) drawn in the half-plane $t > 0$

$$\int_c u\, dx - f(u)\, dt = 0. \tag{1.6}$$

2. *Determination*

One may obtain the weak solutions as follows, (Lax, 1957). Consider the single-valued function defined in $t > 0$ by

$$\phi(x, t) = \int_{0,0}^{x,t} u\, dx - f(u)\, dt, \tag{1.7}$$

and assume first that $u(x, t)$ is a classical solution and that $u_0(x)$ is differentiable. Then

$$\frac{\partial \phi}{\partial t} + f\left(\frac{\partial \phi}{\partial x}\right) = 0. \tag{1.8}$$

The general integral of (1.8) is obtained by eliminating s between

$$\phi = sx - f(s)t - c(s)$$

and

$$0 = x - f'(s)t - c'(s); \tag{1.9}$$

the initial data give, if $\phi_0(x)$ is a primitive of $u_0(x)$,

$$\begin{aligned} \phi_0(x) &= sx - c(s), \\ 0 &= x - c'(s). \end{aligned} \tag{1.10}$$

They allow one to find $c(s)$ for the given initial data. Put $y = c'(s)$, then $s = \phi_0'(y) = u_0(y)$, and the solution of the Cauchy problem is defined by

$$\begin{aligned} \phi &= (x-y)s - f(s)t + \phi_0(y), \\ x - y &= tf'(s). \end{aligned} \tag{1.11}$$

But if $g(\tau)$ is the reciprocal function of $\tau = f'(u)$ and if $G(\tau)$ is the convex function defined by

$$G(\tau) = \tau g(\tau) - f[g(\tau)],$$

whose derivative is $g(\tau)$, one can write (1.11) as

$$\phi = tG\left[\frac{x-y}{t}\right] + \phi_0(y) = F(x, y, t), \tag{1.12}$$

$$\phi_0'(y) = g\left(\frac{x-y}{t}\right). \tag{1.13}$$

Thus one can state (Lax, 1957)

THEOREM 1.1. *For given x and t, the value(s) of the eventually multivalued classical solution $\phi(x, t)$ of* (8) *is one of the extrema of $F(x, y, t)$, when this function is considered as a function of the single variable y.*

We recover of course the previous classical solution of (1.2):

$$u(x, t) = \partial\phi/\partial x = u_0(y),$$

with

$$x - y = tu_0(y).$$

Now, if $u(x, t)$ is a single-valued *weak solution, bounded and continuous in* x *for* t *fixed except for an enumerable set of values of* x, $\phi(x, t)$ is a *continuous* function which satisfies (1.8) almost everywhere and which is defined by (1.12) and (1.13). Then we have

THEOREM 1.2. *$u(x, t)$ is the x derivative of the function $\phi(x, t)$ equal, for a given pair x and t, to one of the extrema of $F(x, y, t)$ when this function is considered as a function of y alone. At a point (x, t) u may be discontinuous if there exist* at least *two different values* of y which give to F the *same extremum value.*

3. *Jump Relation*

Now, if (Γ) is a line in the x, t plane on which $u(x, t)$ is discontinuous, for every sub arc PQ of (Γ) we can write

$$\int_{PQ} [u]\, dx - [f(u)]\, dt = 0 \tag{1.14}$$

if we call $[h]$ the jump of $h(x, t)$ across (Γ):

$$[h] = h(x+0, t) - h(x-0, t). \tag{1.15}$$

At every point of (Γ) where the curve admits a tangent, note by $w = dx/dt$ its slope; then one can state

THEOREM 1.3. *On a line of discontinuity* (Γ) *one may write*:

$$w[u] = [f(u)]. \tag{1.16}$$

Equation (1.16) is the "jump relation" for Eq. (1.2). For Eq. (1.1), the jump relation is simply

$$w = \bar{u} \tag{1.17}$$

if we put

$$\bar{h} = \tfrac{1}{2}(h(x+0, t) + h(x-0, t)). \tag{1.18}$$

Conversely it is easy to prove

THEOREM 1.4. *If $u(x, t)$ is a bounded and piecewise continuous and differentiable function on the x, t halfplane, which satisfies* (1.2) *at every regular point and* (1.16) *on every interior point of regular arcs* (Γ) *where u is discontinuous, then $u(x, t)$ is a weak solution of* (1.2).

In the field of weak solutions, it can be shown that *every Cauchy problem has at least one solution.* But it may in fact have more than one solution. For instance take the example of Fig. 2 for Eq. (1.1); the function

$$u = 0 \quad \text{if} \quad 2x < t,$$
$$u = 1 \quad \text{if} \quad 2x > t,$$

is another weak solution of the same Cauchy problem. Thus, the set (cs) of classical solutions was too narrow and the set of weak solutions (ws) is too broad.

C. Shock Solutions (ss)

To resolve the previous ambiguity, we go back to one of the ideas given in the introduction. We can consider that Eq. (1.1) or (1.2) gives a simplified description of a physical process in which dissipative effects have been neglected. Thus we may try to consider the mathematical model which takes dissipation into account and see what happens when it tends towards zero.

1. *Burgers Equation*

One very famous mathematical model is the following equation

$$\frac{\partial u}{\partial t} + u\frac{\partial u}{\partial x} = \nu\frac{\partial^2 u}{\partial x^2} \tag{1.19}$$

introduced by Burgers (1948) in order to build a model for statistical turbulence. Equation (1.19) has many applications and interpretations in fluid dynamics. See, for instance, Germain (1960a), Hayes (1960), and Lighthill (1956). It has been thoroughly investigated from the mathematical point of view in a fundamental paper by Hopf (1950). We note that for $\nu = 0$, (1.19) is identical with (1.1). In what follows ν is a positive constant (the reason for this choice will be explained below).

Burgers nonlinear equation may be solved explicitly. Namely, according to (1.19), there exists a function $\phi(x, t)$ such that

$$u\,dx - \left(\frac{u^2}{2} - \nu\frac{\partial u}{\partial x}\right)dt = d\phi,$$

and ϕ is a solution of

$$\frac{\partial \phi}{\partial t} + \frac{1}{2}\left(\frac{\partial \phi}{\partial x}\right)^2 = \nu\frac{\partial^2 \phi}{\partial x^2}, \tag{1.20}$$

which for $\nu = 0$ is just Eq. (1.8) when $f(u) = u^2/2$. Now, if we put $\phi = -2\nu \operatorname{Log}\theta$, $\theta(x, t)$ is a solution of the heat equation

$$\partial\theta/\partial t = \nu\partial^2\theta/\partial x^2.$$

Then, the classical solution of this equation shows that the solution of the Cauchy problem for (1.19) is (Hopf, 1950; Cole, 1951)

$$u(x, t) = \int_{-\infty}^{\infty} \frac{x-y}{t} \exp\left\{-\frac{1}{2\nu} F(x, y, t)\right\} dy \Big/ \int_{-\infty}^{\infty} \exp\left\{-\frac{1}{2\nu} F(x, y, t)\right\} dy \tag{1.21}$$

where

$$F(x, y, t) = \frac{(x-y)^2}{2t} + \phi_0(y) \tag{1.22}$$

which is precisely the function F defined by (1.12) for Eq. (1.1).

Now according to Hopf (1950), we can prove

THEOREM 1.5. *For fixed Cauchy data* $(u_0(x), \phi_0(x))$, *when the positive constant* ν *tends towards zero,* $\phi(x, t)$ *tends towards*

$$\phi(x, t) = \operatorname*{Inf}_y \{F(x, y, t)\}, \tag{1.23}$$

which is a continuous function. Let (D) be the set of the $t > 0$ *half plane where the absolute minimum of F is reached for one and only one y—say,* $y_m(x, t)$. *In (D) the limiting value of* (1.21) is

$$u = (x - y_m)/t, \tag{1.24}$$

The complementary set of (D) is cut in an enumerable set by every line $t =$ *const. At these points* $u(x, t)$ *is discontinuous and we have*

$$u(x-0, t) = \frac{x - y_-(x, t)}{t}, \qquad u(x+0, t) = \frac{x - y_+(x, t)}{t},$$

where $y_-(x, t)$ *and* $y_+(x, t)$ *are (for given x, t) respectively the smallest and largest values of y which give to F its absolute minimum value. As a consequence at a point discontinuity*

$$u(x-0, t) > u(x+0, t). \tag{1.25}$$

The solution given by (1.23) will be called the shock solution of (1.1). The reader will find in Hopf's paper the study of the main properties of such a solution, defined as the limit when ν tends towards zero of the Burgers equation (1.19).

2. *Definition of a Shock Solution for Eq.* (1.2)

Following Lax (1957), we can state the definition as a generalization of the preceding result.

DEFINITION. *For fixed Cauchy data* $(u_0(x), \phi_0(x))$, *the shock solution of Eq.* (1.8) *is given by*

$$\phi(x, t) = \operatorname*{Inf}_y \{F(x, y, t)\}, \tag{1.26}$$

where $F(x, y, t)$ *is defined by* (1.12). *At every point* (x, t) *the shock solution of* (1.2) *is given by*

$$u(x-0, t) = g\left(\frac{x-y_-(x, t)}{t}\right), \qquad u(x+0, t) = g\left(\frac{x-y_+(x, t)}{t}\right), \tag{1.27}$$

where g *is the reciprocal function of* f' *and* $y_-(x, t)$ *and* $y_+(x, t)$ *are defined as in Theorem 1.5.* Note that at any regular point $y_-(x, t) = y_+(x, t)$.

It is possible to prove the following results—see Lax (1957).

THEOREM 1.6. 1. *The shock solution is continuously differentiable at every point of an open domain* (D)*: the intersection of a line* $t = \text{const}$, *with the complementary of* (D) *is an enumberable set.*

2. *The shock solution is a weak solution.*

3. *On every arc* (Γ) *of the complementary of* (D) *the jump relation* (1.16) *is satisfied. Moreover*

$$u(x-0, t) \geq u(x+0, t). \tag{1.28}$$

4. *When* $t \to 0$, $\phi(x, t)$ *defined by* (1.26) *tends towards* $\phi_0(x)$. *Then* $u(x, t)$ *converges (weakly) towards* $u_0(x)$.

Thus the shock solution which has been defined gives one (unique) solution of the Cauchy Problem. The following result allows one to assume that this solution is physically sensible.

THEOREM 1.7. *If a family of functions* $u_0^{(n)}(x)$ *converges weakly when* n *goes to infinity towards a function* $u_0(x)$, *the corresponding shock solution*

$$\phi^{(n)}(x, t)$$

of the Cauchy problem converges towards $\phi(x, t)$ *and in* (D) $u^n(x, t)$ *converges towards* $u(x, t)$.

This result shows that the future is continuously determined by the past. But the contrary is false as shown by the example of Fig. 4.

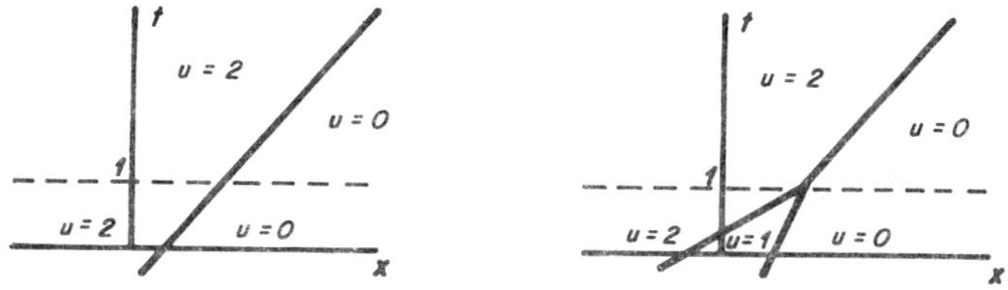

FIG. 4. Equation (1): two different pasts ($t < 1$) may give rise to the same future ($t > 1$).

Now, the time $t=0$ does not play a special role in the definition of a shock solution, as shown by

THEOREM 1.8. *The solution* ϕ *defined by* (1.26), *where* F *is given by* (1.12), *also checks the following identity*

$$\phi(x, t_1+t_2)=\operatorname*{Inf}_y \left\{\phi(y, t_1)+t_2 G\left(\frac{x-y}{t_2}\right)\right\},$$

where t_1 *and* t_2 *are positive.*

3. *Characterizations of Shock Solutions*

First of all, as suggested by Section I,C,1 for Eq. (1.1), it is possible to obtain the shock solutions of (1.2) as a limit of solutions when dissipation is taken into account. This is shown in particular by the following (Ladyzhenskaya, 1956)

THEOREM 1.9. *Let* $u_0^{(n)}(x)$ *be continuous and bounded functions, with continuous and bounded first derivatives, which converge towards* $u_0(x)$; *for* n *infinite, Let* $f_m(u)$ *be increasing functions with two continuous derivatives such that* $f_m''(u), f_m'(u), f_m(u)$ *tend towards* $f''(u), f'(u), f(u)$ *respectively, uniformly in any bounded interval for* n *infinite. Call* $u_{m,n}(x, t)$ *the classical solution of*

$$\frac{\partial u}{\partial t}+\frac{\partial}{\partial x}(f_m(u))=\nu_n\frac{\partial^2 u}{\partial x^2}, \tag{1.29}$$

where ν_n *are positive constants, which take the initial values* $u_0^{(n)}(x)$. *Then,* $u_{m,n}$ *tends towards the shock solution of* (1.2)—*with Cauchy data* $u_0(x)$—*when* m, n *increase indefinitely and* ν_n *tends towards* 0.

But it is also of great interest to indicate which supplementary condition must be added to the definition of a weak solution in order that this weak solution be a shock solution. The following results give, among others, answers to this requirement. The first one proves that inequality (1.28) gives a possible condition (Germain and Bader, 1953);

THEOREM 1.10. *Among all piecewise continuous differentiable functions* $u(x, t)$ *which are weak solutions of* (1.2) *and which take the value* $u_0(x)$ *for* $t=0$, *there is one and only one which checks the inequality.* (1.28).

Roughly speaking, it says that we must supplement the jump relation (1.16) by an inequality, (1.28), in order to define the subset of shock solutions among all the weak solutions. In most usual applications in physical problems, shock solutions are in fact defined by "jump relations" and "jump inequalities." Let us state also a more precise result (Oleinik, 1956):

THEOREM 1.11. *There exists one and only one weak solution of* (1.2) *which checks* (1.5), *such that in every bounded domain* (D) *of the* x, t *half-plane* ($t > 0$) *one may write*

$$\frac{u(x+\Delta, t) - u(x, t)}{\Delta} \leq K(t)$$

if $(x+\Delta, t)$ *and* (x, t) *belong to* (D), Δ *is positive, and if* $K(t)$ *is a decreasing function of* t (*which may take arbitrarily large values when* t *tends towards zero*).

A comment at this point is in order. It must be noted that in any case an inequality is needed in order to select shock solutions among weak solutions, and that is particularly clear in Theorems 1.10 and 1.11. But the question immediately arises: is it possible to find the jump inequality (1.28) when considering only Eq. (1.2)? The answer is in the affirmative; it is sufficient to impose a "stability condition," which says that in the field of shock solutions, Cauchy data must define unambiguously the solution and that this solution must depend continuously on the data. To be more specific, if $u_0(x, t)$ is a solution, consider the perturbed solution $u = u_0(x, t) + \delta\tilde{u}(x, t)$ which takes the initial given values $u_0(x) + \delta\tilde{u}_0(x)$, δ being a small parameter. The data $\tilde{u}_0(x)$ must define *uniquely* $\tilde{u}(x, t)$ and the system of discontinuity lines of $u(x, t)$ must be infinitely close to the corresponding system for $u_0(x, t)$. As an example, let $u_0(x, t)$ be the solution of (1.1) given by

$$u_0(x, t) = a \quad \text{if} \quad x < 0,$$

$$u_0(x, t) = -a \quad \text{if} \quad x > 0,$$

and let us look for the *linear* perturbed solution of Eq. (1.1). Now $\tilde{u}$ is a solution of the linear equation

$$\frac{\partial \tilde{u}}{\partial t} + u_0(x, t)\frac{\partial \tilde{u}}{\partial x} = 0$$

defined by the initial data $\tilde{u}_0(x)$. It must admit a discontinuity line very close to $x = 0$, defined by its slope $\delta\tilde{\omega}(t)$. If

$$\tilde{u}_1(t) = \tilde{u}(-0, t), \qquad \tilde{u}_2(t) = \tilde{u}(+0, t),$$

one has

$$2\tilde{\omega}(t) = \tilde{u}_1(t) + \tilde{u}_2(t).$$

If $a < 0$ (see Fig. 5) the data do not define $\tilde{u}$ uniquely; more precisely $\tilde{u}_1(t)$ and $\tilde{u}_2(t)$ may be chosen arbitrarily. But if a is positive one has

$$\tilde{u}_1(t) = \tilde{u}_0(-at), \qquad \tilde{u}_2(t) = \tilde{u}_0(at),$$

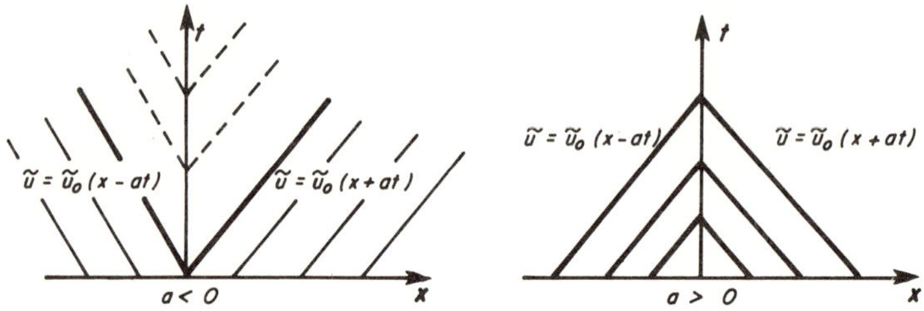

FIG. 5. Perturbation of the initial data of a particular weak solution for deriving the jump inequality.

and the perturbed solution is uniquely defined. Thus for this special example the stability condition implies

$$u(x - 0, t) \geq u(x + 0, t),$$

which is precisely the shock inequality (1.28). In fact it is not difficult to show, with a convenient linear change of variables, that this inequality must be satisfied at every point of a discontinuity line for the stability condition to be fulfilled.

When shock solutions are defined by a limiting process of solutions taking account of a vanishingly small dissipation, the inequality is $\nu > 0$ in Eq. (1.19) or (1.29). This condition may be regarded as arising from the following *stability condition*: *dissipation must have for large time an attenuative effect and not a growing effect.*

Take for instance the solution of (1.1): $u(x, t) = a$. Superpose for $t = 0$ an infinitely small harmonic perturbation

$$u_0(x) = a + \delta \exp (i\omega x).$$

Put $u(x, t) = a + \delta\tilde{u}(x, t)$. The unknown function $\tilde{u}(x, t)$ is a solution of

$$\frac{\partial \tilde{u}}{\partial t} + a \frac{\partial \tilde{u}}{\partial x} = \nu \frac{\partial^2 \tilde{u}}{\partial x^2},$$

and then

$$\tilde{u}(x, t) = \exp \{i\omega(x - at) - \nu\omega^2 t\}.$$

Thus if ν was not positive, dissipation would have a growing effect.

These remarks show that in both cases, the shock inequality may be related to a stability condition for the behavior of perturbations when *time is increasing*. Roughly speaking, this inequality arises only from the privileged evolution of time which goes from the past to the future.

In physical applications, we know a general principle which governs the evolution of systems: the second law of thermodynamics. But we must keep in mind that the application of this second law will give *necessary* shock inequalities; but that we have no assurance that these inequalities will be sufficient to select the correct shock solutions among the field of weak solutions. Methods similar to those given here for the single equation (1.2) must be used.

D. Remarks on the Formulation of a Conservation Law

It may be noted that Eq. (1.2) may be reduced to Eq. (1.1) with a change of dependent variable. For, if $v(x, t)$ is a solution of (1.2), $u = f'(v)$ is a solution of (1.1). As a result, *the set of classical solutions of* (1.2) *is mapped on the set of classical solutions of* (1.1) *by the mapping* $u \to g(u)$.

But it is easy to show that this result cannot be extended to the *field of weak solutions*. In general the transform of a discontinuous solution of (1.2) is not a discontinuous solution of (1.1), as may be seen when looking at the jump relations.

This shows that the usual way to describe a physical system which satisfies some conservation laws by the related partial differential system is not the most convenient one, because the usual classical analysis cannot be applied without restrictions to weak solutions. To avoid such difficulties, we must describe such a physical system by the laws written in integral form, as given by the fundamental laws of physics. For instance (1.2) may be related to the conservation law

$$\frac{\partial}{\partial t}\int_a^b u\,dx + f(u)\Big]_a^b = 0, \tag{1.30}$$

where $f(u)]_a^b$ means $f(u(b)) - f(u(a))$. It says that whatever be the fixed domain $a \leq x \leq b$ the rate of the quantity defined by its linear density u is balanced by the associated flux, defined by $f(u)$. As will be shown more generally in the next section such a formulation implies without ambiguity the partial differential equation (1.2) and the jump relation for the discontinuous solution (1.16).

II. Shock Conditions. Examples from Gas Dynamics and Magneto-Fluid Dynamics

A. General Formulation of a Conservation Law

1. *The Associated Partial Differential Equations*

Generally speaking, a conservation law is one which says that for a given material within a domain $(\mathscr{D})$, bounded by a closed surface (S), which we follow during its motion, the rate of what is "furnished" by the exterior is equal to the rate of what "flows" outside (S) plus the rate of what is "located" inside $(\mathscr{D})$.

Such a statement introduces three associated quantities we have to represent by mathematical entities. To be specific, we use orthogonal Cartesian coordinate and we assume that the quantity for which the balance is written is a second-order tensor. Then we have the volume density of what is "located," say $\mathscr{A}_{ij}$, the volume density of the rate of what is "furnished," say A_{ij}, and the surface density of the rate of what "flows," say a_{ijk}, this last quantity being, in this case, a third-order tensor. Now the precise statement of the conservation law is

$$\frac{d}{dt}\iiint_{(\mathscr{D})} \mathscr{A}_{ij}\, dv + \iint_{(S)} a_{ijk} n_k\, d\sigma = \iiint_{(\mathscr{D})} \mathscr{A}_{ij}\, dv \tag{2.1}$$

if n_i is the exterior normal unit vector and d/dt the material derivative operator.

If the domain $(\mathscr{D})$ is moving independently of the medium, call $\mathbf{W}(P, t)$ the absolute velocity of a point P of (S), $\mathbf{U}(P, t)$ the velocity of the *particle* located at P at time t, $\mathbf{V}(P, t)$ the relative velocities of the medium at P, with respect to the surface (S), $\mathbf{U} = \mathbf{V} + \mathbf{W}$, and $\delta/\delta t$ is the rate of variation with time. One gets from (2.1)

$$\frac{\delta}{\delta t}\iiint_{(\mathscr{D})} \mathscr{A}_{ij}\, dv + \iint_{(S)} a''_{ijk} n_k\, d\sigma = \iiint_{(\mathscr{D})} A_{ij}\, dv, \tag{2.2}$$

where

$$a''_{ijk} = a_{ijk} + \mathscr{A}_{ij} U_k$$

and, in particular, for a domain (absolutely) fixed:

$$\frac{\partial}{\partial t}\iiint_{(\mathscr{D})} \mathscr{A}_{ij}\, dv + \iint_{(S)} a'_{ijk} n_k\, d\sigma = \iiint_{(\mathscr{D})} A_{ij}\, dv, \tag{2.3}$$

with

$$a'_{ijk} = a_{ijk} + \mathscr{A}_{ij} U_k. \tag{2.4}$$

From this general statement one can derive the partial differential equations associated with this law. Take for instance the case of a fixed domain $(\mathscr{D})$, and assume that $\mathscr{A}_{ij}$ and a'_{ijk} are continuously differentiable in a volume (V), $((\mathscr{D}) \subset (V))$. As $(\mathscr{D})$ is an arbitrary subdomain of (V), one may write, by use of the divergence theorem,* in any interior point of (V):

$$\frac{\partial}{\partial t}\mathscr{A}_{ij} + a'_{ijk,k} = A_{ij},$$

or

$$\frac{\partial}{\partial t}\mathscr{A}_{ij} + (\mathscr{A}_{ij} U_k)_{,k} + a_{ijk,k} = A_{ij}. \tag{2.5}$$

* We denote the partial derivative $\partial f/\partial x_k$ by $f_{,k}$ and we use the summation convention.

2. *The Associated Jump Relation*

Now, we assume that the fixed domain ($\mathscr{D}$) is the sum of two subdomains ($\mathscr{D}_1$) and ($\mathscr{D}_2$), adjacent to a common boundary (Σ), located inside ($\mathscr{D}$), having a continuously varying tangent plane at each of its points P. Call $\mathbf{W}(P, t)$, the velocity field of the points of (Σ) and $\mathbf{N}$ the normal unit vector at P to the surface (Σ), pointing inside ($\mathscr{D}_2$). By hypothesis, $\mathscr{A}_{ij}$, a_{ijk}, U_i are continuously differentiable bounded functions inside ($\mathscr{D}_1$) and ($\mathscr{D}_2$), but may be discontinuous when crossing (Σ). Let us call $\mathbf{V}(P, t)$ the relative velocity of the medium with respect to (Σ) and $[f]$, the discontinuity of f, when crossing (Σ) in the direction $\mathbf{N}$.

We will apply the conservation law to a family of domains ($\mathscr{D}_\delta$), defined by a closed subset (Σ_0) of (Σ): ($\mathscr{D}_\delta$) is the set of points M such that $\mathbf{PM} = z\mathbf{N}$, $P \in (\Sigma_0)$, $-\delta \leq z \leq \delta$, δ being a small length; ($\mathscr{D}_{\delta,1}$) and ($\mathscr{D}_{\delta,2}$) are respectively the subdomains of ($\mathscr{D}_\delta$) defined by $-\delta \leq z \leq 0$ and $0 \leq z \leq \delta$; (S_b) is the part of the boundary (S) of ($\mathscr{D}_\delta$) obtained from points P belonging to the boundary of (Σ_0); ($S_{-\delta}$) and (S_δ) are the parts of

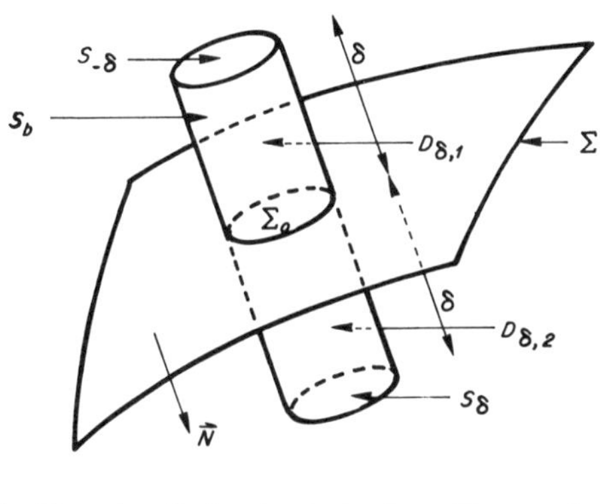

Fig. 6. Domains D_δ for the derivation of jump relations.

(S) defined by $z = -\delta$ and $z = \delta$, respectively. Now the following lemma can be proven

LEMMA. *If* $w = \mathbf{N} \cdot \mathbf{W}$, *one may write*

$$\frac{\partial}{\partial t} \iiint_{(\mathscr{D}\delta)} \mathscr{A}_{ij}\, dv = \iiint_{(\mathscr{D}\delta)} \frac{\partial A_{ij}}{\partial t}\, dv - \iint_{(\Sigma 0)} [A_{ij}]\, w\, d\sigma. \tag{2.6}$$

With this lemma, it is easy to prove

THEOREM 2.1. *In each point of* (Σ), *if* $v = \mathbf{V} \cdot \mathbf{N}$

$$[\mathscr{A}_{ij} v + a_{ijk} N_k] = 0. \tag{2.7}$$

Namely, apply (2.3) to the domain $(\mathscr{D}_\delta)$, taking account of (2.6)

$$\iiint_{(\mathscr{D}_\delta)} \frac{\partial \mathscr{A}_{ij}}{\partial t} dv + \iint_{(S_b)} (\mathscr{A}_{ij} U_k + a_{ijk}) n_k \, d\sigma + \iint_{(S_\delta)} (A_{ij} U_k + a_{ijk}) n_k \, d\sigma$$

$$+ \iint_{(S_{-\delta})} (\mathscr{A}_{ij} U_k + a_{ijk}) n_k \, d\sigma - \iint_{(\Sigma_0)} [\mathscr{A}_{ij}] \, w \, d\sigma - \iiint_{(\mathscr{D}_\delta)} A_{ij} \, dv = 0.$$

By assumption, all the integrands are bounded. Assume that δ tends towards zero. All the volume integrals and the integral on (S_b) tend towards zero; (S_δ) and $(S_{-\delta})$ tend towards (Σ_0); but $\mathbf{n}$ on $(S_{-\delta})$ and $\mathbf{n}$ on (S_δ) tend towards $\mathbf{N}$ and $-\mathbf{N}$ respectively. Thus, at the limit we get

$$\iint_{(\Sigma_0)} [A_{ij} v + a_{ijk} N_k] \, d\sigma = 0.$$

But (Σ_0) is an arbitrary closed subset of (Σ), and the integrand is continuous. Then we have (2.7) at every point of (Σ). Reciprocally, one can state the following result which generalizes Theorem 4 of Chapter 1.

THEOREM 2.2. *In any domain* (V) *where* $\mathscr{A}_{ij}$, A_{ij}, a_{ijk}, U_i *are functions with piecewise continuous and bounded derivatives, which satisfy* (2.5) *at any point of continuity and* (2.7) *on every surface of discontinuity, the conservation law* (2.1) *is valid for every subdomain* $(\mathscr{D})$ *of* (V).

B. CLASSICAL GAS DYNAMICS (INVISCID FLUID, HAYES, 1958, 1960; LIEPMANN AND ROSHKO, 1957)

We assume that the thermodynamic quantities are ruled by the classical results of thermodynamics. The specific internal energy e is a given function of the specific volume* τ and the specific entropy s. Pressure and temperature are defined by the identity

$$de = T\, ds - p\, d\tau. \tag{2.8}$$

In particular if we write

$$p = g(\tau, s) \tag{2.9}$$

we assume that g obeys the Weyl inequalities (Weyl, 1949):

$$g_\tau < 0, \qquad g_{\tau\tau} > 0, \qquad g_s > 0. \tag{2.10}$$

If viscosity and heat conduction are neglected and if no external sources of mass momentum and energy are present, the classical conservation laws

* $\tau = \rho^{-1}$, if ρ is the mass density.

may be written as formula (2.1) according to the scheme in Table 1. The application of (2.5) gives the well known system of partial differential

TABLE 1

	$\mathscr{A}$	a	A
Mass	ρ	0	0
Momentum	ρU_i	$p\,\delta_{ij}$	0
Energy	$\rho(e + \frac{1}{2}U^2)$	pU_i	0

equations of gasdynamics for a perfect fluid. The jump conditions are given by (2.7)

$$[\rho v] = 0; \qquad [\rho v\mathbf{U} + p\mathbf{N}] = 0; \qquad [\rho(e + \tfrac{1}{2}\mathbf{U}^2) + p(v + w)] = 0 \quad (2.11)$$

Thus ρv is continuous across (Σ)—say $m = \rho v$. If $m = 0$, $v_1 = v_2 = 0$; no fluid crosses (Σ); (Σ) is called a *contact discontinuity, p is continuous on* (Σ) (but ρ and T may be discontinuous).

If $m \neq 0$, relation (2.11) may be written in the form

$$[\rho v] = 0, \qquad \text{(conservation of mass)} \qquad (2.12a)$$

$$[p + \rho v^2] = 0, \qquad (2.12b)$$

(conservation of momentum)

$$[\mathbf{V}_\mathrm{T}] = 0, \qquad (2.12c)$$

$$[h + \tfrac{1}{2}v^2] = 0, \qquad \text{(conservation of energy)} \qquad (2.12d)$$

where $\mathbf{A}_\mathrm{T}$ is the tangential component of a vector $\mathbf{A}$ and h is the specific enthalpy.

$$h = e + p\tau. \qquad (2.13)$$

The surface (Σ) is called a *shock wave* and (2.12) are called the shock equations. We note that the shock equations may be written with the relative velocity $\mathbf{V}$ only. Such a result is a consequence of *Galilean relativity*; for the laws of mechanics must have the same expression in any inertial frame. (That means they must be invariant if $\mathbf{W}$ is replaced by $\mathbf{W} + \mathbf{W}_0$ and $\mathbf{U}$ by $\mathbf{U} + \mathbf{W}_0$, where $\mathbf{W}_0$ is a constant velocity.)

C. Classical Magneto-Fluid Dynamics (MFD)

1. *General Equations* (Cowling, 1957)

If the fluid is a perfect conductor, the main electrodynamical quantities we have to take into account are the magnetic induction $\mathbf{B}$ and the electric field $\mathbf{E}$, since in magnetofluid dynamics the displacement current is neglected and charge neutrality is assumed. The two vector fields satisfy Maxwell's laws, which can be written in an integral form

$$\iint_{(\Sigma)} \mathbf{B} \cdot \mathbf{n}\, d\sigma = 0; \qquad \frac{\partial}{\partial t}\iint_{(S)} \mathbf{B} \cdot \mathbf{n}\, d\sigma + \int_{(C)} \mathbf{E} \cdot \boldsymbol{\tau}\, ds = 0. \tag{2.14}$$

Here (Σ) is a closed surface, (S) is an open fixed surface, (C) its boundary, $\mathbf{n}$ a unit vector normal to (S), and $\boldsymbol{\tau}$ the unit vector tangential to (C) directly oriented around $\mathbf{n}$ (as in Stokes' theorem). Now if (S) is moving and $\mathbf{W}(P, t)$ is the velocity of a point P of (S) the second law (2.14) may be written

$$\frac{\delta}{\delta t}\iint_{(S)} (\mathbf{B} \cdot \mathbf{n})\, d\sigma + \int_{(C)} (\mathbf{E} + \mathbf{W} \wedge \mathbf{B}) \cdot \boldsymbol{\tau}\, ds = 0 \tag{2.15}$$

because of the identity

$$\frac{\delta}{\delta t}\iint_{(S)} (\mathbf{A} \cdot \mathbf{n})\, d\sigma = \frac{\partial}{\partial t}\iint_{(S)} (\mathbf{A} \cdot \mathbf{n})\, d\sigma + \iint_{(S)} (\operatorname{div} \mathbf{A})\, \mathbf{W} \cdot \mathbf{N}\, d\sigma$$
$$+ \int_C \operatorname{curl} (\mathbf{A} \wedge \mathbf{W}) \cdot \boldsymbol{\tau}\, dS.$$

In the classical theory, $\mathbf{B}$ is the same in any two Galilean frames (R) and (R'); but $\mathbf{E}$ is not. Assume that the motion of (R') is a translation with a (constant) velocity $\mathbf{W}_0$. According to (2.15), we introduce the quantity

$$\mathbf{E} + \mathbf{U} \wedge \mathbf{B}$$

at a given instant t, and at a point M where the fluid velocity is $\mathbf{U}$, which is by definition the local electric field. Moreover, when the fluid is a *perfect conductor*, i.e. when Joule's dissipation is neglected, this field is identically null. In this case, the only fundamental electrodynamical quantity is the induction $\mathbf{B}$.

In order to express the interaction of the electrodynamical field with matter, we have to introduce the Maxwell Stress Tensor

$$\Pi_{ij} = (1/\mu)B_i B_j - \delta_{ij}(B^2/2\mu), \qquad B^2 = B_i B_i, \tag{2.16}$$

and the Poynting vector

$$\boldsymbol{\Sigma} = (1/\mu)\mathbf{E} \wedge \mathbf{B}. \tag{2.17}$$

Here μ is the magnetic permeability.

With these notations, when all the dissipative effects are neglected (viscosity, heat conduction, Joule's effect), the fundamental equations of (MFD) may be written as conservation laws according to the following scheme presented in Table 2.

TABLE 2

	$\mathscr{A}$	a	A
Maxwell equations	0	B_i	0
	B_i	$-U_i B_j$	0
Mass	ρ	0	0
Momentum	ρU_i	$p\,\delta_{ij} - \Pi_{ij}$	0
Energy	$\rho(e + \frac{1}{2}U^2) + \dfrac{B^2}{2\mu}$	$pU_i + \Sigma_i$	0

Then by application of Theorem 2.1, it is possible to write the jump equations on a surface of discontinuity. According to Galilean invariance, these equations may be written in a form in which only the relative velocity $\mathbf{V}$ (and not $\mathbf{U}$) appears. Easy calculations show that they can be expressed in the following form, if we note $B_0 = \mathbf{B} \cdot \mathbf{N}$,

$$[B_0] = 0, \qquad \text{(Maxwell's equations)} \tag{2.18a}$$

$$[v\mathbf{B}_T] - B_0[\mathbf{V}_T] = 0, \tag{2.18b}$$

$$[\rho v] = 0, \qquad \text{(conservation of mass)} \tag{2.18c}$$

$$\left[p + \rho v^2 + \frac{B_T{}^2}{2\mu}\right] = 0, \tag{2.18d}$$

$$\left[\rho v \mathbf{V}_T - \frac{B_0}{\mu}\mathbf{B}_T\right] = 0, \qquad \text{(conservation of momentum)} \tag{2.18e}$$

$$\rho v\left[h + \frac{v^2}{2}\right] + \left[\frac{v B_T{}^2}{\mu}\right] - \frac{B_0}{\mu}[\mathbf{B}_T \cdot \mathbf{V}_T] = 0. \qquad \text{(conservation of energy)} \tag{2.18f}$$

Note that (2.18b) says that the tangential component of $\mathbf{B} \wedge \mathbf{V}$ is continuous. This may be proved directly, starting from (2.15), by reasoning in a very similar way as in Section II,A, above.

2. *Classification of Discontinuities* (Bazer and Erickson, 1959; Germain, 1959; Shercliff, 1960; Anderson, 1963)

First of all $m = \rho v$ is continuous on (Σ).

(a) $m = 0$: (Σ) is a contact discontinuity; $v_1 = v_2 = 0$; the flow does not cross (Σ).

TABLE 3

Classification of MFD Discontinuities

		$[\mathbf{B}_T]$	$[\mathbf{V}_T]$	$[p]$	$[\tau]$	Jump relations	Nature of the discontinuity
$m=0$	$B_0 \neq 0$	0	0	0			Th-CD
$v=0$	$B_0=0$					$\left[p+\frac{B^2}{2\mu}\right]=0$	MFD-CD
$m>0$	$B_0=0$	0	0	0	0		No discontinuity
$[v]=0$	$B_0 \neq 0$			0	0	$\begin{cases} v^2=\frac{B_0{}^2}{\mu\rho}=\alpha_n{}^2 \\ \frac{\mathbf{V}_T}{v}=\frac{\mathbf{B}_T{}'}{B_0} \\ [B_T{}^2]=0 \end{cases}$	A-SW
$m>0$, $[v]\neq 0$	$B_0=0$		0			$\begin{cases} [v\mathbf{B}_T]=0 \\ \left[p+\frac{B_T{}^2}{2\mu}+m^2\tau\right]=0 \\ \left[h+\frac{\tau B_T{}^2}{2\mu}+\frac{m^2\tau^2}{2}\right]=0 \end{cases}$	Normal MFD-SW
	$B_0 \neq 0$					See Eq. (2.21)	Oblique MFD-SW

(i) If $B_0 \neq 0$, (2.18b) and (2.18e) show that $\mathbf{V}_T$ and $\mathbf{B}_T$ are continuous and (2.18d) that p is continuous. The only possible discontinuities are those of the thermodynamic quantities which are not functions of p alone. We say that it is a purely thermodynamical contact discontinuity (ThCD).

(ii) If $B_0=0$, $[\mathbf{V}_T]$ and $[\mathbf{B}_T]$ may take arbitrary values; the only jump condition which is not trivial (2.18d) says that $p+(B^2/2\mu)$ is continuous. Every thermodynamic quantity may be discontinuous. It is called an MFD contact discontinuity (MFD-CD).

(b) $m \neq 0$; (Σ) *is a shock wave*. Two main cases have to be considered:

(i) *If v is continuous*, B_0 is not zero if (Σ) is *really* a surface of discontinuity. First $[\tau]=0$; then according to (2.18b) and (2.18e)

$$v^2=B_0{}^2/\mu\rho=\alpha_n{}^2 \tag{2.19}$$

and

$$[\mathbf{V}_T] = \frac{v}{B_0}[\mathbf{B}_T] = \frac{B_0}{\mu\rho v}[\mathbf{B}_T], \tag{2.20}$$

where α_n is the *normal Alfvén velocity*; (2.18d) and (2.18f) lead, after some calculations, to the result

$$[p + (B_T^2/2\mu)] = 0, \qquad m[h] + (v/2\mu)[B_T^2] = 0,$$

and then, after elimination of p, $[e]=0$, $[B_T^2]=0$. As a result, *all the thermodynamic quantities are continuous*, and the magnitude of $\mathbf{B}_T$ and of $\mathbf{V}_T$ is continuous. The two vectors $\mathbf{B}$ and $\mathbf{V}$ are subject to a rotation around $\mathbf{N}$, one of them being arbitrary. In such a case, (Σ) is called a *rotation shock* or an *Alfvén shock wave* (A-SW).

(ii) If $[v] \neq 0$, (Σ) is an *MFD shock wave*. If $B_0 = 0$ (normal MFD-SW) $[v\mathbf{B}_T]$ and $\mathbf{V}_T$ are continuous (2.18b, 2.18e); then the shock conditions are reduced to

$$\left[\left(p + \frac{B_T^2}{2\mu}\right) + m^2\tau\right] = \left[\left(h + \frac{B_T^2}{\mu}\tau\right) + \frac{m^2\tau^2}{2}\right] = 0,$$

and this case presents some analogy with the case of classical gas dynamics.

In the general case when $B_0 \neq 0$ (oblique MFD-SW), we will reformulate conditions (2.18) in a convenient way for future applications. First we introduce an orthonormed frame $(\mathbf{N}, Py_2, Py_3)$ at every point P, and note by A_2 and A_3 the components of a vector $\mathbf{A}$ on Py_2 and Py_3. Then we remark that a condition $[X]=0$ means that there exists a constant a such that equations $X_1 - a = 0$, $X_2 - a = 0$ are valid on both faces of the shock. Therefore we have the following (Germain, 1959, 1960b):

THEOREM 2.3. *At every point P of a shock wave (Σ), it is possible to define some constants B_0, E_2, E_3, m, P_1, P_2, P_3, C in such a way that on both sides of the shock one has*

$$\begin{aligned}
&\mathscr{L}_1 = (m\mu)^{-1}\{mB_2\tau - B_0V_2 + E_3\} = 0,\\
&\mathscr{L}_2 = V_2 - (m\mu)^{-1}B_0B_2 - P_2 = 0,\\
&\mathscr{L}_3 = p + m^2\tau + (2\mu)^{-1}[B_2^2 + B_3^2] - P = 0,\\
&\mathscr{L}_4 = h + \tfrac{1}{2}m^2\tau^2 + (2\mu)^{-1}[V_2^2 + V_3^2] + (m\mu)^{-1}[E_2B_3 - E_3B_2] - C = 0,\\
&\mathscr{L}_5 = (m\mu)^{-1}\{mB_3\tau - B_0V_3 - E_2\} = 0,\\
&\mathscr{L}_6 = V_3 - (m\mu)^{-1}B_0B_3 - P_3 = 0.
\end{aligned} \tag{2.21}$$

B_0 is the normal magnetic induction; E_2 and E_3 define the tangential electric field; m is the mass flux; P_1, P_2, P_3 are the momentum flux; and C is the energy flux. These quantities are *the "constants" of the shock* (at P).

Now, if we call $q_1, q_2, \ldots, q_6$ the variables $B_2, V_2, \tau, T, B_3, V_3$, and $\mathscr{E}_6$ the space of the variables q_i, to find a shock in MFD one must look for pairs of points in $\mathscr{E}_6$ which for *given shock constants* obey the equations (2.21): $\mathscr{L}_p = 0$ ($p = 1, 2, \ldots, 6$).

Remark. If one *excludes* the case of the *rotation shock*, one sees according to $\mathscr{L}_1 = \mathscr{L}_2 = \mathscr{L}_5 = \mathscr{L}_6 = 0$ that two particular cases may be considered:

1. $v_1 = \alpha_{n1}$, or $\tau_1 = \tau_* = (m\mu)^{-1}B_0^2$ and $(\mathbf{B}_T)_2 = 0$; that is the "switch off" shock.
2. $v_2 = \alpha_{n2}$, or $\tau_2 = \tau_*$, and $(\mathbf{B}_T)_1 = 0$, that is the "switch on" shock.

Now if τ_1 and τ_2 are different from τ_*, either $(\mathbf{B}_T)_1 = (\mathbf{B}_T)_2 = 0$, or $B_3 B_2^{-1}$ is constant through the shock. Then,

THEOREM 2.4. *Except in the case of a rotation shock, the normal* $\mathbf{N}$ *and the upstream and downstream magnetic inductions* $(\mathbf{B})_1$ *and* $(\mathbf{B})_2$ *are coplanar.*

The axes Py_2 and Py_3 may be chosen in order to have $B_3 = 0$ on both sides of the shock; then V_3 is continuous. One may take $V_3 = 0$ when relating the motion in a new frame (translation; velocity V_3 along Py_3). Then $P_3 = E_2 = 0$ and $\mathscr{L}_5$ and $\mathscr{L}_6$ are identically zero. Another simplification is possible (with a convenient translation along Py_2). One may take $E_3 = 0$, then in this frame $\mathbf{V}$ and $\mathbf{B}$ are collinear on both sides of the shock, or $P_2 = 0$. This will sometimes be used in what follows and referred to as the *simplified formulation of shock conditions.* When using this formulation, we will write

$$E_3 = m\mu E_*, \qquad V = V_2,\ B = B_2, \qquad B_0^* = (m\mu)^{-1}B_0,$$

and the simplified shock conditions will be

$$\begin{aligned}
\mathscr{L}_1 &= \mu^{-1}B\tau - B_0^* V + E_* = 0,\\
\mathscr{L}_2 &= V - B_0^* B = 0,\\
\mathscr{L}_3 &= p + m^2\tau + (2\mu)^{-1}B^2 - P = 0,\\
\mathscr{L}_4 &= h + \tfrac{1}{2}(m^2\tau^2 + V^2) - E_* B - C = 0.
\end{aligned} \tag{2.22}$$

D. PROPERTIES OF SHOCKS IN MFD

1. *Shock Generating Function* (Germain, 1959, 1960b)

Let us formally introduce the function $\mathscr{P}$ defined by

$$\mathscr{P} = \frac{1}{T}\left(\frac{m^2\tau^2}{2} + \frac{V_2^2 + V_3^2}{2} + \frac{B_2^2 + B_3^2}{2\mu}\tau + \frac{E_3 B_2 - E_2 B_3}{m\mu} - f(\tau, T)\right.$$
$$\left. - \frac{B_0}{m\mu}(B_2 V_2 + B_3 V_3) + C - P_1\tau - P_2 V_2 - P_3 V_3\right), \tag{2.23}$$

where f is the specific free energy of the fluid

$$f = e - Ts = h - p\tau - Ts. \tag{2.24}$$

The interpretation of this function will be given in the next Section. Then if we consider the q_i as *the* independent variables, i.e. the function $\mathscr{P}$ $(B_2, V_2, \tau, T, B_3, V_3)$ we can check the following identities (see Eqs. 2.21):

$$\mathscr{L}_1 = T\partial\mathscr{P}/\partial B_2, \qquad \mathscr{L}_2 = T\,\partial\mathscr{P}/\partial V_2, \qquad \mathscr{L}_3 = T\,\partial\mathscr{P}/\partial\tau,$$

$$\mathscr{L}_4 = T^2\frac{\partial\mathscr{P}}{\partial T} + T\left[\tau\frac{\partial\mathscr{P}}{\partial\tau} + V_2\frac{\partial\mathscr{P}}{\partial V_2} + V_3\frac{\partial\mathscr{P}}{\partial V_3}\right],$$

$$\mathscr{L}_5 = T\,\partial\mathscr{P}/\partial B_3, \qquad \mathscr{L}_6 = T\,\partial\mathscr{P}/\partial V_3.$$

Then we can state

THEOREM 2.5. *For given values of the shock constants, two states connected through a shock have images in the q_i space $\mathscr{E}_6$—say S_a and S_b—which are stationary points of the function $\mathscr{P}(q_i)$.*

Incidentally, let us note that $s = \mathscr{P} + T(\partial\mathscr{P}/\partial T)$. Then for given shock constants, the study of the "possible" shocks is the study of the stationary points of the function $\mathscr{P}(q_i)$. This function $\mathscr{P}$ is called a *shock generating function.*

In the simplified formulation (see Eqs. 2.22) the function $\mathscr{P}$ is given by

$$\mathscr{P} = \frac{1}{T}\left(\frac{m^2\tau^2}{2} + \frac{V^2}{2} + \tau\frac{B^2}{2\mu} + E_* B - B_0{}^* BV + C - P\tau - f\right), \tag{2.25}$$

and the study of stationary points s of $\mathscr{P}$ has to be done in the $\mathscr{E}_4$ subspace (q_1, q_2, q_3, q_4).

2. *Existence and Properties of Points S*

Following a method introduced by Weyl (1949), we consider in this study the properties of the function $\mathscr{P}$ *around a point S* when the constants of the shock are continuously varying. For instance, differentiation of (2.25) shows that in such conditions

$$T\,d\mathscr{P} = T\,ds = dC - \tau\,dP - BV\,dB_0{}^* + B\,dE_* + \tfrac{1}{2}\tau^2\,dm^2.$$

It is not possible to give here all the details of the proofs which lead to the following results. They are important for the study of MFD problems, but we prefer here to concentrate on those aspects of the theory which may have a general value. Proofs may be found in the literature (Germain, 1959,

1960b; Anderson, 1963), and some of these results appear as special cases of a general theory of shocks, such as the one given by Lax (1957).

Let us define for a given state the slow and fast wave speeds, a and A, as the positive roots of the equation

$$P(x) = x^4 - (c^2 + \alpha_n{}^2 + \alpha_T{}^2)x^2 + \alpha_n{}^2 c^2 = 0; \tag{2.26}$$

c is the usual sound speed $c^2 = \partial p(\rho, s)/\partial \rho$ while α_n and α_T are the normal and tangential Alfvén speeds—see (2.19). It may be shown that a, α_n, $A (a \leq \alpha_n \leq A)$ are the (local) three speeds of propagation of infinitesimal perturbations. Then it may be shown:

THEOREM 2.6. *For given values of the shock constants, there exist at most* 4 *points* S, S_1, S_2, S_3, S_4, *the order of the index corresponding to nonincreasing values of* τ *and to nondecreasing values of* s. *At these points, the following identities are valid*:

$$S_1: A \leq v, \qquad S_2: \alpha_n \leq v \leq A,$$
$$S_3: a \leq v \leq \alpha_n, \qquad S_4: v \leq a.$$

Through a shock, the discontinuities of p, P, T, s *have the same sign.* It will be shown later that all these discontinuities are in fact positive. In classical gas dynamics, $a = \alpha_n = 0$ and $A = c$. Only the points S_1 and S_2 have to be considered. In MFD we have $a \leq c \leq A$.

E. Conclusion

The purpose of this section was essentially to show how to obtain the shock conditions from a given set of conservation laws for nondissipative systems, to apply the general formulas to specific classical examples taken from classical gas dynamics and magnetofluid dynamics, and to state for these examples the kind of general results of physical significance which may be obtained. Our main goal is not too discuss deeply all the properties but mainly, as was said in the Introduction, to show what is a shock when considering a hierarchy of mathematical models of increasing physical complexity. That leads us to the study of the shock structure which is the topic of the next section.

III. Classical Shock Structure

In this Section more elaborate mathematical models will be considered. In particular we expect, according to the simple example of Burgers' equation, that the discontinuities which appear in the simplified version will be smoothed in those models. In order to start with an example, we will consider first the MFD equation when dissipative phenomena are not neglected.

A. Equations of MFD (with Dissipative Effects)

1. *General Equations*

When viscous effects are taken into account, the stress tensor will be written

$$\sigma_{ij} = -p\,\delta_{ij} + \tau_{ij}$$

with τ_{ij} the viscous stress (symmetric) tensor.

When *heat conduction* is taken into account, we have to introduce the heat flux vector $\tilde{\omega}_i$. If $\mathbf{n}$ is the outward normal unit vector to the boundary (S) of some closed domain $(\mathscr{D})$, the rate of heat received by $(\mathscr{D})$ is the surface integral of $-\tilde{\omega}_i n_i$. When *electrical dissipation* is taken into account, the local electric field

$$\hat{\mathbf{E}} = \mathbf{E} + \mathbf{U} \wedge \mathbf{B} \tag{3.1}$$

is not zero.

As far as conservation of mass momentum and energy are concerned, they are written schematically as in Table 4.

TABLE 4

	$\mathscr{A}$	a	A
Mass	ρ	0	0
Momentum	ρU_i	$p\,\delta_{ij} - \Pi_{ij} - \tau_{ij}$	0
Energy	$\rho(e + \frac{1}{2}U^2) + \dfrac{B^2}{2\mu}$	$PU_i - \tau_{ij}U_j + \Sigma_i + \tilde{\omega}_i$	0

The five scalar equations so obtained must be supplemented by the Maxwell equations

$$\frac{\partial \mathbf{B}}{\partial t} + \operatorname{curl} \mathbf{E} = 0, \tag{3.2}$$

$$\operatorname{div} \mathbf{B} = 0. \tag{3.3}$$

Moreover, conservation of charge in MFD allows one to write the current density $\mathbf{J}$ as the curl of some vector $\mathbf{H}$, the magnetic field, and the classical assumption of linearity and isotropy of the medium from the electromagnetic point of view leads us to write $\mathbf{B} = \mu \mathbf{H}$, that is to say,

$$\mu \mathbf{J} = \text{curl}\, \mathbf{B}, \tag{3.4}$$

which allows one in this case to compute $\mathbf{J}$ when $\mathbf{B}$ is known. Let us call *principal dependent variables* the eight quantities p, T, U_i, B_i and *principal equations* the five equations in Table 4 and the three equations (3.2), and note that Equation (3.3) is a consequence of (3.2) if it is only assumed that the induction field is initially divergence free. The various thermodynamic quantities are supposed to be defined by the state equation and $\mathbf{J}$ by (3.4). The variables τ_{ij}, E_i, $\tilde{\omega}_i$ will be called secondary. Thus in order to determine a flow, we must write twelve other scalar equations, which will be the dissipation laws of the system.

2. *Dissipation Laws* (Kulikovskii, 1962a)

A general process often used in such cases is to write the *rate of production of entropy*. It is easy to confirm that the equations in Table 4 may be written as

$$\rho \frac{d\tau}{dt} - U_{i,i} = 0, \tag{3.5}$$

$$\rho \frac{dU_i}{dt} + p_{,i} - \Pi_{ij,j} - \tau_{ij,j} = 0, \tag{3.6}$$

$$\rho \frac{d}{dt}\left(e + \frac{U_2}{2}\right) + pU_{i,i} + p_{,i} U_i - \tau_{ij,j} U_i - \tau_{ij} U_{j,i} + \frac{\partial}{\partial t}\left(\frac{B^2}{2\mu}\right) + \Sigma_{i,i} + \tilde{\omega}_{i,i} = 0, \tag{3.7}$$

and to check also the following identity

$$\frac{\partial}{\partial t}\left(\frac{B^2}{2\mu}\right) + \Sigma_{i,i} + \hat{E}_i J_i + U_i \Pi_{ij,j} = 0. \tag{3.8}$$

Then using (3.5)–(3.8), with the factors p, $-U_i$, 1, -1, respectively, one gets

$$\rho T \frac{ds}{dt} - \tau_{ij} U_{j,i} - \hat{E}_i J_i + \tilde{\omega}_{i,i} = 0, \tag{3.9}$$

or

$$\frac{\partial}{\partial t}(\rho s) + \left(\rho U_i s + \frac{\tilde{\omega}_i}{T}\right)_{,i} = \sigma^* = \frac{1}{T}\left[\tau_{ij} U_{i,j} + \hat{E}_i J_i - \frac{\tilde{\omega}_i T_{,i}}{T}\right].$$

which says that for any closed domain $(\mathscr{D})$

$$\frac{d}{dt}\iiint_{(\mathscr{D})} \rho s \, dv + \iint_{(S)} \frac{\tilde{\omega}_i}{T} n_i \, d\sigma = \iiint_{(\mathscr{D})} \sigma^* \, dv,$$

which would be interpreted as a conservation law if σ^* was known. According to the second law of thermodynamics, σ^* is the *volumetric rate of production of entropy* because the second term of the left-hand side is the opposite of $-\iint_S (Q/T)\, d\sigma$ if Q is the surface density of the heat rate received by $(\mathscr{D})$ in conduction through (S).

First it can be proved that if no dissipation mechanisms are found

$$\tau_{ij} = 0, \qquad \hat{E}_i = 0, \qquad \tilde{\omega}_i = 0,$$

which is precisely the case considered in the previous chapter. Then if ε_{ij} is the strain rate tensor,

$$T\sigma^* = \tau_{ij}\varepsilon_{ij} + J_i \hat{E}_i - \tilde{\omega}_i (\log T)_{,i}; \tag{3.10}$$

$\tau_{ij}, J_i, -\tilde{\omega}_i$ are called generalized fluxes, and ε_{ij}, $\hat{E}_i$, $(\log T)_{,i}$ generalized forces. The laws of *thermodynamics of irreversible phenomena state that every generalized flux is a linear combination of the generalized forces of the same tensorial structure.*

Assume first *isotropy*; τ_{ij} is a linear function of ε_{kl}, J_i and $\tilde{\omega}_i$ are linear functions of $\hat{E}_i$ and $(\log T)_{,i}$, and $T\sigma^*$ is a quadratic form of the generalized forces—or of the generalized fluxes—and such a quadratic form must be *nonnegative.*

One can write more explicitly when the dissipative effects are uncoupled:

$$\tau_{ij} = (\eta_1 - 2\eta_2)\varepsilon_{kl}\delta_{ij} + 2\eta_2 \varepsilon_{ij}, \tag{3.11}$$

$$\hat{E}_i = \sigma^{-1} J_i, \tag{3.12}$$

$$\tilde{\omega}_i = -kT_{,i}; \tag{3.13}$$

η_1 and η_2 are the coefficients of viscosity (η_2 is the classic one, $3\eta_1 - 4\eta_2$ is the bulk viscosity), σ is the electrical conductivity—in the expression (3.10) for $T\sigma^*$ there appears in this case the classical Joule effect $\sigma^{-1}J^2$, and k is the heat conduction coefficient. In order for $T\sigma^*$ not to be negative, we must have

$$\eta_2 \geq 0, \qquad 3\eta_1 - 4\eta_2 \geq 0, \qquad \sigma \geq 0, \qquad k \geq 0.$$

Of course, all these coefficients of dissipation may depend on the principal dependent variables. In fact, they may practically depend on T and B^2. Let us note simply that (3.12) may be written

$$\mathbf{J} = \sigma(\mathbf{E} + \mathbf{U} \times \mathbf{B}), \tag{3.14}$$

which is the classical Ohm's law when the conductivity is scalar. Of course, the classical gas dynamic equations are obtained by putting $\Pi_{ij}=0$, $\Sigma_i=0$ in Table 4 and by taking only (3.11) and (3.13) for the dissipation laws.

B. Shock Structure Concept

Let us recall—see the Introduction—that we are essentially concerned with the study of systems with small dissipative coefficients (in MFD, $\eta_1, \eta_2, \sigma^{-1}$, and k will be assumed to be small); and especially with the local behavior of the solutions in the neighborhood of a surface (Σ) which is a shock when these coefficients are exactly zero. More precisely let us consider the flow, defined by some *fixed* conditions, of a family of fluids of vanishingly small dissipative coefficients. Assume that a shock (Σ) is present in the flow in the limiting case when the fluid is perfect; what is the local behavior of the flow in the neighborhood of (Σ) when the coefficients are small but not zero? This question will be fully discussed at least from the point of view of classical gas dynamics in Section IV, where the complete asymptotic expansion will be given. One of the conclusions is that the "first term" of this local expansion may be found if one *looks for a stationary flow* in which all the physical quantities depend only one one single variable, say x, the abscissa along the normal of Σ being conveniently magnified. In such a flow, values at $x=-\infty$ and at $x=+\infty$ of the variable quantities correspond to the values taken on both sides of the shock in the *perfect* fluid flow. This gives a reasonable proof of what was assumed in the classical literature; it is why this solution is called the *classical shock structure solution* or, with the terminology of Section IV, the *zero order shock structure.* This section will be devoted to this kind of flow: *steady solutions with all the dependent variable functions of the single independent variable x.* We want to exhibit in various cases the mathematical problems arising with these solutions, their profiles, especially when the dissipative coefficients tend towards zero, and sketch some of the physical conclusions from this study.

C. Classical Magneto-Fluid Dynamics (Germain 1959, 1960b)

1. *General Results*

We are then looking for a stationary flow in which all the quantities depend on $x_1=x$ only. Thus from (3.3) $B_1=\text{const}=B_0$; from (3.2), $E_2=\text{const.}$, $E_3=\text{const.}$; from (3.4), $J_1=0$. Conservation laws in Table 4 may be integrated once. If $U_1 = u(u > 0)$, $\rho u = m$ and $u = m\tau$ where m is a constant. Finally, if we eliminate E_1, J_2, J_3 we obtain

$$
\begin{aligned}
(\mu\sigma)^{-1}\, dB_2/dx &= mB_2\tau - B_0 V_2 + E_3, \\
\eta_2\, dV_2/dx &= m\{V_2 - (m\mu)^{-1} B_0 B_2 - P_2\}, \\
\eta_1\, du/dx &= p + m^2\tau + (2\mu)^{-1}(B_2^2 + B_3^2) - P, \qquad (3.15)\\
k\frac{dT}{dx} &= m\left(h + \frac{m^2\tau^2}{2} + \frac{V_2^2 + V_3^2}{2}\right) + \frac{E_2 B_3 - E_3 B_2}{\mu} \\
&\quad - \eta_1 u \frac{du}{dx} - \eta_2 V_2 \frac{dV_2}{dx} - \eta_2 V_3 \frac{dV_3}{dx} - mC, \\
(\mu\sigma)^{-1}\, dB_3/dx &= mB_3\tau - B_0 V_3 - E_2, \\
\eta_2\, dV_3/dx &= m\{V_3 - (m\mu)^{-1} B_0 B_3 - P_3\}.
\end{aligned}
$$

Here P, P_2, P_2 and C are constants of integration for momentum and energy. Let us introduce the variables q_i and the shock generating function $\mathscr{P}(q_i)$, as in Section II. Note $\dot{q}_i = dq_i/dx$; and $\mathscr{D}(\dot{q}_i)$ is the volumetric rate of production of entropy σ^*:

$$
\begin{aligned}
\mathscr{D}(\dot{q}_i) = \frac{1}{T}\Bigg\{&(\sigma\mu^2)^{-1}\left[\left(\frac{dB_2}{dx}\right)^2 + \left(\frac{dB_3}{dx}\right)^2\right] + m^2\eta_1\left(\frac{d\tau}{dx}\right)^2 \\
&+ \eta_2\left[\left(\frac{dV_2}{dx}\right)^2 + \left(\frac{dV_3}{dx}\right)^2\right] + \frac{k}{T}\left(\frac{dT}{dx}\right)^2\Bigg\}.
\end{aligned} \qquad (3.16)
$$

Then it is not difficult to prove:

THEOREM 3.1. *The Shock structure equations may be written in the compact form*

$$\partial\mathscr{D}/\partial\dot{q}_i = 2m\, \partial\mathscr{P}/\partial q_i. \qquad (3.17)$$

A flow corresponds to an integral curve of (3.18) in $(\mathscr{E}_6)$. Since $\mathscr{D}$ is a quadratic form we have

THEOREM 3.2. *Along an integral curve,*

$$m(\mathscr{P}(x_2) - \mathscr{P}(x_1)) = \int_{x_1}^{x_2} \mathscr{D}\, dx. \qquad (3.18)$$

and as $\mathscr{D}$ is a positive quadratic form,

THEOREM 3.3. *$\mathscr{P}(x)$ is an increasing function of x.*

A very important point is that we look for a bounded solution (BS) defined for every x, $\mathscr{P}(q_i)$ is bounded. By (3.18), $\mathscr{D}(x)$ must tend towards zero with $|x|^{-1}$; the same is true for each $\dot{q}_i$ and therefore for $\partial\mathscr{D}/\partial\dot{q}_i$ (a linear function of $\dot{q}_i$), and for $\partial\mathscr{P}/\partial q_i$ (Eq. 3.17). Then when x goes from $-\infty$ to $+\infty$, the integral curve goes from one stationary point to another, i.e. it connects two points S (see Theorem 2.5), say S_I and S_F; as

$$s = \mathscr{P} + T(\partial\mathscr{P}/\partial T)$$

one has

THEOREM 3.4. *There exist at most four points S which may be end points of a solution (BS) of* (3.17) *and which are the singular points of this system. If an integral curve connects two of these points* S_I *and* S_F, *one has* $s(S_I) \leq s(S_F)$.

As in Section II, with a change of frame, one may assume $V_3 = B_3 = 0$ for such a solution and, in this simplified formulation, use the shock generating function (2.25).

2. *Singular Points*

As we look for an integral curve (3.5) joining two singular points (S), it is necessary to investigate the nature of the solutions of (3.17) in the neighborhood of a point S. Call $(q_j)_0$ its coordinate, and put $q_j = (q_j)_0 + \bar{q}_j$. According to a classical result it is sufficient to investigate the behavior of the system

$$\partial\bar{\mathscr{D}}/\partial\dot{\bar{q}}_i = 2m\, \partial\bar{\mathscr{P}}/\partial\bar{q}_i, \tag{3.19}$$

where $\mathscr{D}(\dot{\bar{q}}_i)$ and $\bar{\mathscr{P}}(\bar{q}_i)$ are quadratic forms of the principal increments of $\mathscr{D}(\dot{q}_i)$ and $\mathscr{P}(q_i)$ in the neighborhood of S.

$$\mathscr{D}(q_j, \dot{q}_i) = \mathscr{D}((q_j)_0, \dot{\bar{q}}_i) + \cdots = \bar{\mathscr{D}}(\dot{\bar{q}}_i) + \cdots,$$
$$\mathscr{P}(q_j) = \mathscr{P}((q_j)_0) + \bar{\mathscr{P}}(\bar{q}_j) + \cdots, \ \bar{\mathscr{P}} = P_{ij}\bar{q}_i\bar{q}_j,$$

the P_{ij} are elements of a symmetric matrix.

The eigensolutions of (3.19) are proportional to exp (λx), where λ is a (*real*) eigenvalue of a symmetric matrix X_{ij}, simply derived from P_{ij}. The number of positive λ at a point S is equal to the *positive index* of the quadratic form $X_{ij}\xi_j\xi_i$. i.e. of the quadratic form $\bar{\mathscr{P}}$, that is to say to the number of "squares" with a positive coefficient in any decomposition of $\bar{\mathscr{P}}$ in "squares." If we work with the simplified formula (2.25), it can be shown that

$$\bar{\mathscr{P}} = \frac{1}{T_0}\left\{\frac{\mu}{2\tau_0}\overline{\left(\frac{B\tau}{\mu} - B_0{}^*V + E_*\right)}^2 + \frac{1}{2}\left(1 - \frac{\tau_*}{\tau}\right)\overline{\left(V + \frac{\mu B_0{}^* E_*}{\tau - \tau_*}\right)}^2\right.$$
$$\left. + \frac{1}{2}\left(\frac{\partial\sigma}{2T_0}\right)_0\left(\bar{T} - \frac{(\partial s/\partial\tau)_0}{(\partial s/\partial T)_0}\bar{\tau}\right)^2 + \frac{1}{2}\frac{P(u_0)}{\tau_0{}^2(u_0{}^2 - \alpha_n{}^2)}\bar{\tau}^2\right\} \tag{3.20}$$

where $s(\tau, T)$ is the specific entropy and $P(x)$ the polynomial defined by (2.26). Thus, according to Theorem (2.6), as $(1/T_0)(\partial s/\partial T)_0$ is positive because it is the specific heat at constant volume, we have

THEOREM 3.5. *If λ_1, λ_2, λ_3, λ_4 are at a point S, the four eigenvalues in a nondecreasing order, at S_1: $0 \leq \lambda_1 \leq \lambda_2 \leq \lambda_3 \leq \lambda_4$; at S_2 and S_3: $\lambda_1 \leq 0 \leq \lambda_2 \leq \lambda_3 \leq \lambda_4$; at S_4: $\lambda_1 \leq \lambda_2 \leq 0 \leq \lambda_3 \leq \lambda_4$.*

In other words, at a point S where there exist p positive eigenvalues, if the integral curves are oriented in the $\mathscr{P}$ increasing direction, the integral curves which "start" from S form a manifold of dimension p, and the integral curves which "end" at S form a manifold of dimension 4-p.

D. GENERALIZATION TO LINEAR DISSIPATION MECHANISMS (Kulikovskii, 1962a)

In such a case (3.10) may be written

$$T\sigma^* = K_i \dot{q}_i \tag{3.21}$$

with

$$K_i = mTL_{ij}\dot{q}_j, \tag{3.22}$$

where L_{ij} is an element of a (not necessarily symmetric) matrix which depends on the q_k. It will be assumed that $\det(L_{ij}) \neq 0$, and also that $L_{ij}\dot{q}_i\dot{q}_j$ is a nonnegative quadratic form. It can be checked that the flow is ruled by the system

$$\partial\mathscr{P}/\partial q_i = L_{ij}\dot{q}_j \tag{3.23}$$

and that

$$\mathscr{P}(x_2) - \mathscr{P}(x_1) = \int_{x_1}^{x_2} L_{ij}\dot{q}_i\dot{q}_j \, dx. \tag{3.24}$$

Then $\mathscr{P}$ is still an increasing function of x and Theorem 3.4 is still valid.

As far as the behavior of integral curves near a point S is concerned the associated linear system is

$$(L_{ij})_0\dot{\bar{q}}_j = P_{ij}\bar{q}_j.$$

Let us note that

$$(L_{ij})_0 = A_{ij} + B_{ij}, \tag{3.25}$$

A_{ij} being a symmetric matrix, B_{ij} an antisymmetric one, and A_{ij} is positively defined. The eigensolutions are proportional to $\exp\{(\lambda + i\lambda')x\}$ where λ and λ' are real. The integral curve does not lie necessarily in $\mathscr{E}_4$

anymore. But it may be proven that at every point S the number of eigenvalues with a positive *real* part is equal to the positive index of $\mathscr{P}$. As a consequence, the integral curves which may start from (S_1, S_2 and S_3, S_4) of $\mathscr{E}_6$ form a manifold of (4, 3, 2) dimensions respectively; the integral curves which may end at (S_1, S_2 and S_3, S_4) form a manifold of (0, 1, 2) dimensions respectively.

E. Further Generalization (Kulikovskii, 1962b)

All that was said previously may be generalized to every physical one-dimensional system which is governed by equations which can be put into the form

$$\frac{\partial}{\partial t}\left(\frac{\partial \mathscr{A}}{\partial q_i}\right) + \frac{\partial}{\partial \bar{x}}\left(\frac{\partial \mathscr{B}}{\partial q_i}\right) = \frac{\partial}{\partial \bar{x}}\left(L_{ij}\frac{\partial q_j}{\partial \bar{x}}\right), \tag{3.26}$$

where $\mathscr{A}(q_i)$ and $\mathscr{B}(q_i)$ are given scalar functions of $q_i (i = 1, 2, \ldots, n)$ and $L_{ij}(q_j)$ is the matrix of dissipation coefficients. We will assume also that

$$A_{ij} X_i X_j = \frac{\partial^2 \mathscr{A}}{\partial q_i \, \partial q_j} X_i X_j$$

is a positive definite quadratic form of the X_i at every point q_i. *First if we neglect the dissipation* ($L_{ij} = 0$), the shock conditions are easily to be found; if w is the speed of the shock, we obtain

$$\frac{\partial \mathscr{B}}{\partial q_i}(q_k^{(1)}) - w \frac{\partial \mathscr{A}}{\partial q_i}(q_k^{(1)}) = \frac{\partial \mathscr{B}}{\partial q_i}(q_k^{(2)}) - w \frac{\partial \mathscr{A}}{\partial q_i}(q_k^{(2)}), = C_i$$

where C_i and w are the *constants* of the shock. For such a phenomenon a *shock generating function* is obtained by

$$\mathscr{P} = w\mathscr{A} - \mathscr{B} + C_j q_j . \tag{3.27}$$

Two points q_i which may be connected through a shock with given shock *constants* are stationary points of the function $\mathscr{P}(q_i)$.

The discussion of the number of pairs of such points S and their properties is outside the scope of this paper. It may be shown how the classical cases of gas dynamics and magneto-fluid dynamics may be reduced to this mathematical scheme; see for instance (Kulikovskii, 1962b). *Now if we take dissipation into account*, we must write some inequalities in order that the corresponding terms are in fact dissipative terms. These will be obtained as follows.* For any linearized system associated with (3.26), we require

* This is a generalization of what was done in Section I.

that a solution which is periodic in $\bar{x}$ for $t=0$ must remain uniformly bounded when t tends towards infinity. Such a system is

$$(A_{jk})_0 \frac{\partial \bar{q}_k}{\partial t} + (B_{jk})_0 \frac{\partial \bar{q}_k}{\partial \bar{x}} = (L_{jk})_0 \frac{\partial^2 q_k}{\partial \bar{x}^2}.$$

And it is easy to check that for such a condition to be fulfilled it is necessary and sufficient that at every point q_i *the quadratic form* $L_{ij}X_iX_j$ *be nonnegative.*

In order to study the structure of the shock, whose speed is w, we look at *stationary* solutions in the frame defined by the new space variable x

$$x = \bar{x} - wt.$$

Then they are solutions of the system

$$\partial \mathscr{P} / \partial q_i = L_{ij} \dot{q}_j; \tag{3.28}$$

the constants of integration C_i appear in $\mathscr{P}$. See Eq. (3.27). It is clear now that the conclusions of the previous paragraph may be extended to this case.

F. Existence, Uniqueness, Stability

1. *The Fast Shock in MFD* (Germain, 1959, 1960b)

We want to discuss for the simplest case (Section III,C) the existence of an integral arc of the system (3.15) or (3.17) connecting S_1 and S_2. The corresponding solution is by definition the structure of the fast shock. We here sketch the main steps of the proof. The simplified formulation is used.

Lemma 3.1. *There exists an open simply connected domain* $(\mathscr{D})$ *in the space* $\mathscr{E}_4(B, v, \tau, T)$ *in which the following inequalities are valid:*

$$\partial \mathscr{P}/\partial B > 0, \qquad \partial \mathscr{P}/\partial V > 0, \qquad \partial \mathscr{P}/\partial \tau < 0, \qquad \partial \mathscr{P}/\partial T > 0.$$

Moreover S_1 *and* S_2 *belong to the boundary* $(\partial\mathscr{D})$ *of* $(\mathscr{D})$. Such a domain is in the region $\tau > \tau_*$, $T > 0$ and $B(\tau - \tau_*)$ has always the same sign in $(\mathscr{D})$.

Lemma 3.2. *At a point of* $(\partial\mathscr{D})$ *the integral arc of* (3.15), *oriented in the direction of increasing* x, *points towards the exterior of* $(\mathscr{D})$.

The proofs of these two lemmas rest on easy calculations. Now we can state:

Theorem 3.6. *There exists in* $\mathscr{E}_4$ *one and only one integral arc* (L) *in* $(\mathscr{D})$ *whose end points are* S_1 *and* S_2.

Let (F) be the set of points M of the boundary $(\partial\mathscr{D})$ in which $\mathscr{P}(M) > \mathscr{P}(S_1) + \delta$, $\delta > 0$, fixed and sufficiently small. Let (G) be the set of points

N of $(\mathscr{D})$ in which $\mathscr{P}(N) = \mathscr{P}(S_1) + \delta$. The integral curves oriented with positive x give a one to one continuous mapping of (G) on a subset (F') of (F). Every curve starting from N oriented with *decreasing* x goes to S_1 for $x = -\infty$ (Lemma 3.2); the result is obtained if S_2 belongs to (F'). But the boundary of G is identically mapped on the boundary of (F). Then S_2 must be an interior point of (F'); otherwise (F') will have other boundary points which is impossible, because (G) is connex and thus (F') is also connex.

There is no other solution connecting S_1 to S_2 than the one considered in Theorem 3.6, because, according to Theorem 3.5, there is only one direction Δ along which an integral curve may arrive at S_2. If another integral curve existed, $\mathscr{P}(x)$ would not increase from S_1 to S_2. Incidentally the following interesting properties may also be proved:

THEOREM 3.7. *Along this arc (L), $B(x)$, $V(x)$, $T(x)$ $p(x)$, $\rho(x)$ are strictly increasing functions of x.*

Of course in the *physical space* if $q_i(x)$ is a solution, $q_i(x + x_0)$ is also a solution. In order to avoid this nonuniqueness we will choose the origin at the point where $2\tau = \tau_1 + \tau_2$.

The last question and the most important one from the point of view developed in the introduction is what happens to this solution when all the dissipation coefficients η_1, η_2, $(\sigma\mu^2)^{-1}$, k tend towards zero in any manner we can imagine. The answer is given by the following:

THEOREM 3.8. *If one considers a family of flows for values of coefficients of dissipations which go to zero, these flows tend towards the uniform state S_1 for $x < 0$ and the uniform state S_2 for $x > 0$, the convergence being uniform in the complementary of an open neighborhood of $x = 0$.*

Then the solution, valid for the simplified model (no dissipation), given by the state S_1 for $x < 0$ and the state S_2 for $x > 0$ is *always* recovered as the result of any passage to the limit applied to the corresponding shock structure solution when the dissipation coefficients tend towards zero. We say that *the fast shock is stable with respect to the dissipative mechanisms.*

The proof rests with the identity

$$\frac{d\mathscr{P}}{dx} = mT\left\{\sigma\mu^2\left(\frac{\partial\mathscr{P}}{\partial B}\right)^2 + \frac{1}{\eta_2}\left(\frac{\partial\mathscr{P}}{\partial V}\right)^2 + \frac{1}{\eta_1 m^2}\left(\frac{\partial\mathscr{P}}{\partial\tau}\right)^2 + \frac{T}{k}\left(\frac{\partial\mathscr{P}}{\partial T}\right)^2\right\}$$
$$= \mathscr{D}^*.$$

Then, if Q_1 and Q_2 are two points of the integral curve,

$$\mathscr{P}(S_2) - \mathscr{P}(S_1) > \mathscr{P}(Q_2) - \mathscr{P}(Q_1) = \int_{x(Q_1)}^{x(Q_2)} \mathscr{D}^*\, dx. \tag{3.29}$$

Let ε be an arbitrarily small positive number, $(\mathscr{D}_\varepsilon)$ the subdomain of $(\mathscr{D})$, exterior to cubes with centers at S_1 and S_2 and side 2ε, within $(\mathscr{D}_\varepsilon)$ there exists $\varepsilon'(\varepsilon)$ such that

$$mT\left\{\left(\frac{\partial\mathscr{P}}{\partial B}\right)^2+\left(\frac{\partial\mathscr{P}}{\partial V}\right)^2+\frac{1}{m^2}\left(\frac{\partial\mathscr{P}}{\partial\tau}\right)^2+T\left(\frac{\partial\mathscr{P}}{\partial T}\right)^2\right\}>\varepsilon'.$$

If Q_1 and Q_2 are chosen inside $(\mathscr{D}_\varepsilon)$, such that $x(Q_1)<0$, $x(Q_2)>0$, at any point of the arc $Q_1\,Q_2$, we have

$$\mathscr{D}^*>\varepsilon'/\alpha,$$

where α is the upper bound of the dissipation coefficients. Then (3.29) shows that

$$\mathscr{P}(S_2)-\mathscr{P}(S_1)>(\varepsilon'/\alpha)[x(Q_2)-x(Q_1)],$$

or

$$x(Q_2)-x(Q_1)<(\alpha/\varepsilon')(\mathscr{P}(S_2)-\mathscr{P}(S_1))=\delta.$$

Then if ε and δ are given, we may find α in such a way that $B(x)$, $V(x)$, $\tau(x)$, and $T(x)$ differ at most by ε from the values of these quantities, at S_1 if $x\leq-\delta$ and at S_2 if $x\geq\delta$. The theorem is thus proved.

The previous study shows that for any physical quantity—for instance, for the pressure $p(x)$—the graph may be sketched as in Fig. 7. A rough

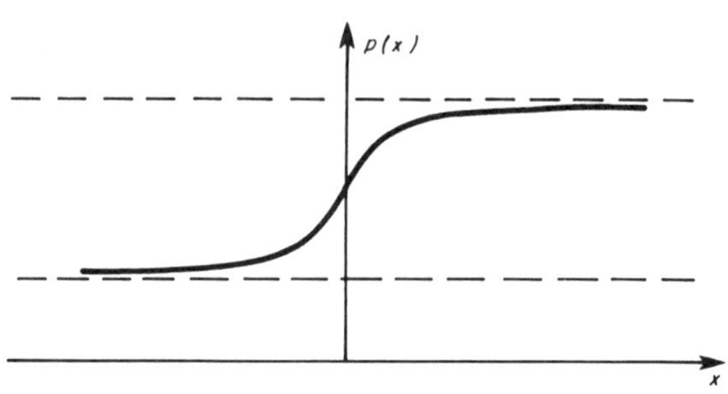

FIG. 7. Variation of a physical quantity; for instance, the pressure inside a shock.

estimate of the shock thickness may be obtained by looking at (3.18). If we replace $(dq_i/dx)^2$ by $[q_i]^2/\delta^2$, we get the estimate

$$\delta=\frac{1}{m[s]}\left\{\left(\frac{\overline{\sigma\mu^2}}{T}\right)^{-1}[B]^2+\left(\frac{\overline{\eta_1}}{T}\right)[v]^2+\left(\frac{\overline{\eta_2}}{T}\right)[V]^2+\left(\frac{\overline{k}}{T_2}\right)[T]^2\right\},\quad(3.30)$$

where $\bar{a}$ is some "rough" mean value of a. The right-hand side of (3.30) is the sum of four lengths, each one formed as the product of a quantity which does not depend on the dissipation and another proportional to the coefficient of dissipation. They can be interpreted as "dissipation lengths" or "diffusion lengths"; and a length is a measure of the importance of the dissipation. Then the thickness of the shock is ruled by the mechanism whose "length" is the greatest. This mechanism is in fact the first which will go into action, when the gradients dq_i/dx grow, in order to avoid the infinite steepening of the shock.

2. *Instability of Intermediate Shocks in MFD* (Germain, 1959)

An intermediate shock joins S_1 or S_2 to S_3 or S_4. We will show that in each of these cases the flow formed by the two constant states across a plane $x = 0$, which is a correct solution for a nondissipative fluid, cannot be the limit of a shock structure solution when the dissipative coefficients tend towards zero, in any way and independently, all the physical quantities remaining uniformly bounded. This result can be forseen by noting that with the simplified formulation the condition

$$B(\tau - \tau_*) + \mu E_* = 0$$

would be satisfied at the limit. Assuming $E_* \neq 0$, this means that B would be infinite in the neighborhood of $\tau = \tau_*$. But we look for *uniformly bounded solutions*; thus difficulties may arise. In fact:

$$\frac{\partial \mathscr{P}}{\partial B} + B_0{}^* \frac{\partial \mathscr{P}}{\partial V} = (\mu T)^{-1}\{B(\tau - \tau_*) + \mu E_*\}.$$

Put $\delta = \mu |E_*|/2B_M$, where B_M is a bound for B. For $\tau_* - \delta \leq \tau \leq \tau_* + \delta$, by application of the Schwarz formula, one may write

$$mT\left\{\left(\frac{\partial \mathscr{P}}{\partial B}\right)^2 + \left(\frac{\partial \mathscr{P}}{\partial V}\right)^2\right\} \geq \frac{K}{T_M}, \tag{3.31}$$

where K is a function of the *shock constants* only. Now take an interval $x_1 \leq x \leq x_2$ such that $\tau(x_1) = \tau_* - \delta$, $\tau(x_2) = \tau_* + \delta$, $\tau_* - \delta \leq \tau(x) \leq \tau_* + \delta$. We find again by application of (3.29) and (3.31)

$$x_2 - x_1 < K_1 \sup[(\sigma\mu^2)^{-1}, \eta_2].$$

On the other hand, the third equation (3.17) shows that

$$|d\tau/dx| < K_2/\inf \eta_1,$$

with K_2 independent, as is K_1, of the dissipation coefficients. Then

$$2\delta < \frac{K^2}{\inf(\eta_1)}(x_2 - x_1) < K_1 K_2 \frac{\sup[(\sigma\mu^2)^{-1}, \eta_2]}{\operatorname{Inf}(\eta_1)}. \tag{3.32}$$

For dissipation mechanisms such that η_2/η_1 or $(\sigma\mu^2)^{-1}/\eta_1$ are small enough the inequality (32) is not satisfied. Then the assumption that B, V, τ, T are uniformly bounded when all the dissipation coefficients tend towards zero is incorrect.

As a consequence *an intermediate shock in a perfect fluid flow model cannot be considered, because it cannot be obtained as the limit for all more elaborate models of the same physical phenomenon when dissipation is taken into account.* Briefly it may be stated that intermediate shocks are *unstable.*

3. *Further Remarks for Other Cases*

For the slow shock in MFD, or for the general case of the system (3.23) or (3.28), no general and complete results have been obtained. The study of the existence of a solution may be approached by an analysis of the topological properties of the domains $\mathscr{P}(q_i) \geq G$, where G is a given number (Kulikovskii, 1962b). By this method it is possible to prove the existence of the shock structure if the shock is not "too strong." That is surely an attractive method for this kind of general investigation. Existence of the slow shock has also been proven for classical MFD in some particular situations.

As far as *stability* is concerned for the system (3.23), the method used above is easily extended. If Λ_{ij} is the inverse matrix of L_{ij} ($\Lambda_{ki} L_{ij} = \delta_{kj}$) the system is written

$$\dot{q}_k = \Lambda_{ki}\, \partial\mathscr{P}/\partial q_i$$

and consequently

$$\frac{d\mathscr{P}}{dx} = \Lambda_{ki} \frac{\partial\mathscr{P}}{\partial q_i} \frac{\partial\mathscr{P}}{\partial q_k},$$

which may be integrated in an interval (x_1, x_2) as in (3.29). It may be shown (Kulikovskii and Liubimov, 1962) that the discussion depends on the behavior of

$$\frac{L_{ij} z_i z_j}{(z_i z_i)^{1/2} (L_{ip} z_p L_{iq} z_q)^{1/2}}. \tag{3.33}$$

When z is kept fixed, the dissipation coefficients L_{ij} tend towards zero. In particular, stability is obtained when (3.33) is uniformly bounded from below by a positive number.

G. Subshocks

We have noted that a length may be associated with each of the dissipative mechanisms, and that some properties connected with the shock structure, as for instance the thickness, are in fact ruled by the mechanism which corresponds to the greatest length. If these lengths are not of the same order of magnitude it may be quite interesting to introduce between the "simplest" model (no dissipation) and the elaborate model (all dissipations present) an *intermediate* model in which only the dominant dissipation mechanisms are taken into account.*

This intermediate model is still a singular perturbation of the simplest one. Such shocks must have a structure because some dissipation is present, but inside this structure some jumps may be present: these jumps are called *subshocks* on *inner shocks* or "*shock within a shock.*"

1. *General Notions* (Germain, 1959)

Let us take a phenomenon governed by n unknown functions q_i of a single independent variable $\bar{x}$. Among these unknowns we consider the secondary variables $q_1(\bar{x}), \ldots, q_r(\bar{x})$, $0 < r < n$, and the principal variables $q_{r+1}(\bar{x}), \ldots, q_n(\bar{x})$. We assume that these variables are solutions of the system

$$\left.\begin{aligned} \varepsilon\eta\,\frac{dq_i}{d\bar{x}} &= F_i(q_j), \\ \eta\,\frac{dq_k}{d\bar{x}} &= F_k(q_j), \end{aligned}\right\} \qquad \begin{aligned} &i = 1, 2, \ldots, r, \\ &j = 1, 2, \ldots, n. \\ &k = r+1, \ldots, n. \end{aligned} \tag{3.34}$$

η is a (small) dissipation coefficient which affects the principal variables; ε is also a small parameter and $\varepsilon\eta$ is the dissipation coefficient which affects the secondary variables. We assume that the basic solution of the simplest model ($\eta = 0$) presents a jump at $\bar{x} = 0$.

In order to study the structure of the shock at $\bar{x} = 0$, we introduce—as it will be shown in the next chapter—a new variable x,

$$\bar{x} = \eta x,$$

based on the greatest dissipation length which is $O(\eta)$; (3.34) becomes

$$\varepsilon dq_i/dx = F_i(q_j), \qquad dq_k/dx = F_k(q_j). \tag{3.35}$$

Now, as ε is small, (3.35) *appears as a* (*singular*) *perturbation of the system*:

$$0 = F_i(q_j), \qquad dq_k/dx = F_k(q_j), \tag{3.36}$$

($i = 1, 2, \ldots, r$, $k = V + 1, \ldots, n$).

If it is possible to solve the r first equations for $q_1, q_2, \ldots, q_r$, (3.36) may be reduced to a system which contains only the *principal variables*. But the solution of this system may have a jump at $x = 0$. In order to study this

* Examples in MFD are given in Marshall (1955), Burgers (1957), Ludford (1959).

possibility and the corresponding jump conditions, we use the method which will be fully explained in the next section. We write (3.35) with a new independent variable ξ,

$$x = \varepsilon\xi,$$

based on the length $\varepsilon\eta$. We get

$$d\dot{q}_i/d\xi = F_i(q_j), \qquad dq_k/d\xi = \varepsilon F_k(q_j). \tag{3.37}$$

The structure of the subshock (if it exists) is defined by the system obtained by putting $\varepsilon = 0$ in (3.37)—that is to say:

$$q_k = q_k^{(0)} = \text{const.}, \qquad k = r+1, \ldots, n,$$
$$dq_i/d\xi = F_i(q_1, q_2, \ldots, q_r, q_{r+1}^{(0)}, \ldots, q_n^{(0)}) \qquad i = 1, 2, \ldots, r. \tag{3.38}$$

Finally two states may be joined by a subshock, if they are defined by the secondary variables $q_1^{(1)}, \ldots, q_r^{(1)}$ and $q_1^{(2)}, \ldots, q_r^{(2)}$ and by the *same* principal variables $q_{r+1}^{(0)}, \ldots, q_n^{(0)}$, and if these two states may be joined by an integral of (3.38) and correspond to the values of the q_i obtained for $\xi = -\infty$ and $\xi = +\infty$, respectively. Then the "jump conditions" of the subshock may be written

$$F_i(q_1^{(1)}, \ldots, q_r^{(1)}, q_{r+1}^{(0)}, \ldots, q_n^{(0)}) = F_i(q_1^{(2)}, \ldots, q_r^{(2)}, q_{r+1}^{(0)}, \ldots, q_n^{(0)}) = 0. \tag{3.39}$$

Thus we can state:

THEOREM 3.9. *In a subshock, the principal variables remain constant and the secondary variables satisfy the jump conditions* (3.39). *The structure of the subshock is governed by* (3.38).

2. *Examples*

1. In classical gas dynamics the shock structure system is

$$\begin{aligned} \eta_1 m_2\, d\tau/dx &= m(p + m^2\tau - P), \\ k\, dT/dx &= m(e - \tfrac{1}{2}m^2\tau^2 + P\tau - c). \end{aligned} \tag{3.40}$$

Assume $\eta_1 = \varepsilon O(k) = \eta$—a small Prandtl number. A subshock is possible for a small Prandtl number; T is the principal variable. Thus, in a subshock, $T = T_0$. The jump condition is $p + m^2\tau - P = 0$. It may be shown that to get a subshock, we must have at S_2: $v = m\tau < c_n < c$ where c_n is the Newtonian sound speed given by $c_n{}^2 = (\partial p/\partial \rho)_T$. Figure 8 shows in the τ, T plane the shape of the integral curves in such a case ($k \neq 0$, $\eta \neq 0$) and the limiting case one gets in subshock ($\eta \to 0$).

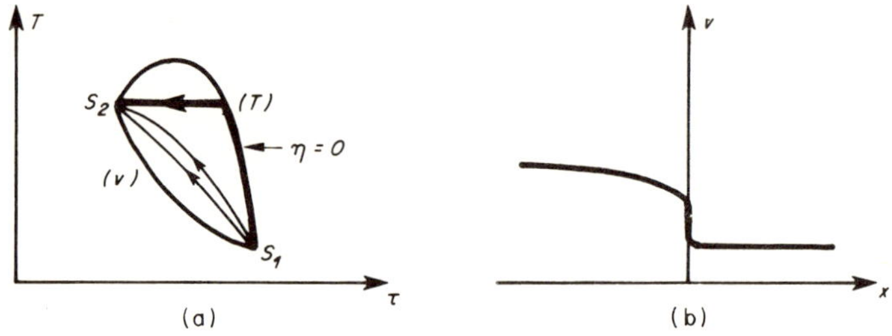

FIG. 8. Thermal subshock when $\eta = 0$. (a) (T) is the line $\partial\mathscr{P}/\partial T = 0$; (v) the line $\partial\mathscr{P}/\partial\tau = 0$. (b) Variation of v in a; thermal subshock.

2. In MFD with a purely transverse magnetic field ($B_0 = 0$) and if only Joule's dissipation is taken into account, the differential system may be written

$$(\sigma\mu^2)^{-1}\frac{dB}{dx} = m\left(\frac{B\tau}{\mu} + E_*\right) = \frac{\partial\mathscr{P}}{\partial B},$$

$$0 = p + m^2\tau + \frac{B^2}{2\mu} - P,$$

$$0 = e - \frac{m^2\tau^2}{2} - \frac{B^2\tau}{2\mu} - E_*B + P\tau - C.$$

The last two equations give a relation between τ and B which is represented by the curve C in Fig. 9. In a subshock B remains constant and the jump conditions are

$$p + m^2\tau = P', \qquad h + \tfrac{1}{2}m^2\tau^2 = C',$$

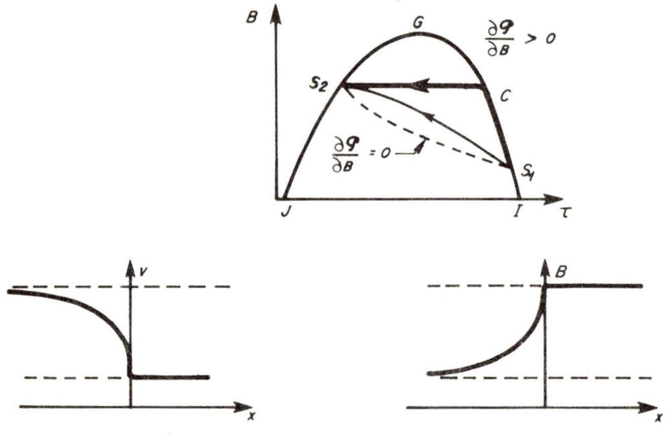

FIG. 9. Ohmic subshock in a transverse magnetic field.

i.e. exactly the usual shock conditions in classical gas dynamics. In order to get a subshock, S_2 must lie on the line GJ of C. That is effectively the case if, at S_2, $v = m\tau < c$, where c is the usual sound speed $c^2 = (\partial p/\partial \rho)_s$. See Fig. 9.

H. Shocks in a Model of a Two-Fluid Plasma

In this section we have shown, taking mostly examples from classical gas dynamics and magneto-fluid dynamics, how discontinuities appear as a simplified version of solutions which may be obtained from a more elaborate scheme—here by the introduction of dissipation effects. One by-product of such a study is to check the consistency of the solution of the simplified scheme. It is why, for instance, we have been led to reject intermediate shocks in MFD.

Such a study of possible refinements of a theory is particularly important in the neighborhood of manifolds along which some peculiar behavior is found. The methods sketched in the present section may be in many cases extended to further problems and provide some interesting insight into the behavior of the physical phenomena.

As a new example, let us note that classical MFD is a rough model for plasma physics. A more refined model will for instance introduce two fluids (ions and electrons) (Germain, 1967). Then with convenient assumptions a one-dimensional flow may be ruled by the following systems if viscosity effects and heat conductions are neglected:

$$v\, dB_2/dx = \xi_2,$$

$$v\, dB_3/dx = \xi_3,$$

$$\beta v \frac{d\xi_3}{dx} = -E_2 + \left(v - \frac{B_0{}^2}{m\mu}\right) B_3 + \chi B_0 \xi_2 - (\sigma\mu)^{-1} \frac{\xi_3}{v},$$

$$\beta v \frac{\partial \xi_2}{dx} = E_3 + \left(v - \frac{B_0{}^2}{m\mu}\right) B_2 - \chi B_0 \xi_3 - (\sigma\mu)^{-1} \frac{\xi_2}{v},$$

$$0 = p + mv + \frac{1}{2\mu}(B_2{}^2 + B_3{}^2) - P,$$

$$0 = m\left(h + \frac{v^2}{2} + \frac{B_0{}^2}{2m\mu^2}(B_2{}^2 + B_3{}^2) + \frac{\beta}{2\mu}(\xi_2{}^2 + \xi_3{}^2)\right.$$

$$\left. - \frac{1}{\mu}(E_3 B_2 - E_2 B_3) - mC. \right.$$

Here m, B_0, E_2, E_3, P, C are constants of the shock; $v = m\tau$; and β and χ are two new coefficients. First, β is $\bar{m}_1\bar{m}_2/\mu\bar{e}^2m$, where $\bar{m}_1$ and $\bar{m}_2$ are the masses of an ion and of an electron and $\bar{e}$ is the charge of the ion and the electron; β is the two-fluid parameter. Second, $\chi = (\bar{m}_1 - \bar{m}_2)/\mu\bar{e}m$ is the Hall parameter.

This model is a refinement of the MFD scheme. If $\beta = \chi = 0$, it reduces to a case which has been examined previously. If $\beta = 0$, $\chi \neq 0$ we get a system valid for a single fluid theory with Hall effect. If $\beta \neq 0$ we have a more elaborate model to schematize a plasma.

We do not want to go into the study (Germain, 1967) of this system, but we would like to mention that the following results can be proved.

1. The specific entropy s (q_i) is a shock generating function of the system (if q_1, q_2, q_3, q_4 *are the quantities* B_2, B_3, ξ_2, ξ_3).

2. The system may be written

$$\mu m T\, \partial s/\partial q_i = L_{ij}\dot{q}_j, \qquad i, j = 1, 2, 3, 4,$$

where L_{ij} is the matrix of dissipation: $L_{ij} = A_{ij} + B_{ij}$

$$A_{ij} = \begin{pmatrix} (\sigma\mu)^{-1} & 0 & 0 & 0 \\ 0 & (\sigma\mu)^{-1} & 0 & 0 \\ 0 & 0 & 0 & 0 \\ 0 & 0 & 0 & 0 \end{pmatrix}, \qquad B_{ij} = \begin{pmatrix} 0 & \chi B_0 v & \beta v & 0 \\ -\chi B_0 v & 0 & 0 & \beta v \\ -\beta v & 0 & 0 & 0 \\ 0 & -\beta v & 0 & 0 \end{pmatrix}.$$

3. Along an integral curve, s is a nondecreasing function.

4. Subshocks are possible. The subshock conditions are identical with the classical shock relations in gas dynamics.

We may apply to this case the results obtained in Section III,D. The presence of B_{ij} gives to the structure some new behavior which does not appear when this gyroscopic matrix is absent. In particular wave trains may appear in the flow upstream or downstream.

The entire study is easy to do in a purely transverse magnetic field (with $B_0 = 0$) (because $\xi_3 = 0$, $B_3 = 0$ for a convenient choice of the frame). For instance, for $(\sigma\mu)^{-1} = 0$, we can achieve in this model a stationary model for a "collisionless shock." Figure 10 shows the typical results which may be obtained and which illustrate the following (Germain, 1967; Morton, 1962):

THEOREM: *If* $(\sigma\mu)^{-1} = 0$, *there exists one and only one solution defined for every* χ, *which tends towards the* S_1 *state for* $x \to -\infty$. When $x \to \infty$ *the solution is discontinuous and tends towards the* S_2 *state if the flow at* S_2 *is subsonic. If the flow at* S_2 *is supersonic, the solution may either be continuous and tend towards the* S_1 *state for* $x \to +\infty$, *or be discontinuous and oscillate periodically around the* S_2 *state.*

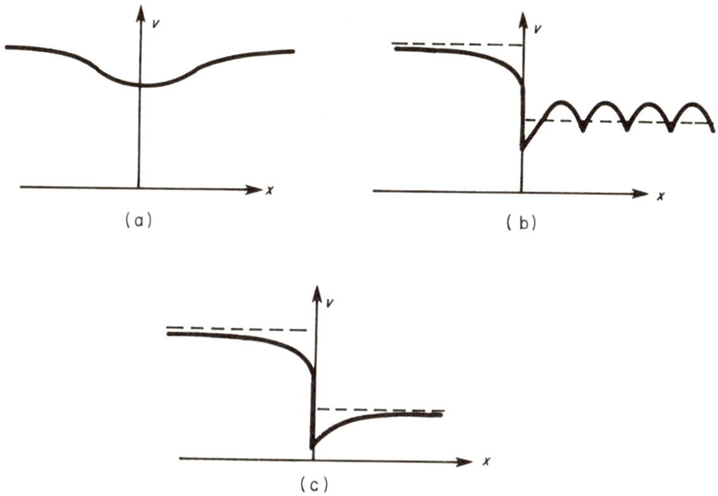

FIG. 10. (a) and (b): Possible variations of $v(x)$ when S_2 is supersonic; (c) variations of $v(x)$ when S_2 is supersonic.

Of course, if Joule dissipation is present, the oscillations are attenuated or may even disappear.

In the general case with a longitudinal component B_0 different from zero, it may happen that oscillations are present not only on the back side of the shock but on the front side also (Morton, 1964).

IV. General Theory of Shock Conditions and Shock Structure in Classical Gas Dynamics

A. Introduction

The present study will be restricted to the classical gas dynamics case. But the methods are in fact general. We consider a flow of a viscous and heat conducting fluid, and the limit of the flow when the dissipative effects tend towards zero. We will then introduce a small parameter $\varepsilon = \mu_0/\rho_0 v_0 L$ where μ_0, ρ_0, v_0, L are typical values for viscosity, volumic mass, velocity, and length, respectively, ε being the inverse of a Reynolds number. Consequently we will write the dissipation laws in dyadic notation:

$$\begin{aligned} \boldsymbol{\tau} &= \varepsilon\mu\{(\nabla \mathbf{U}) + (\nabla \mathbf{U})^T\} + \varepsilon\lambda(\nabla \cdot \mathbf{U})\mathbf{I}, \\ \mathbf{q} &= -\varepsilon k \nabla T, \end{aligned} \tag{4.1}$$

where $\boldsymbol{\tau}$ is the stress tensor, $\mathbf{q}$ the heat flow vector, and $\mathbf{I}$ the unit second-order tensor. The Navier-Stokes equations are written (with convenient dimensionless quantities)

$$\frac{\partial \rho}{\partial t} + \nabla \cdot (\rho \mathbf{U}) = 0,$$

$$\frac{\partial \rho \mathbf{U}}{\partial t} + \nabla \cdot (p\mathbf{I} + \rho \mathbf{U}\mathbf{U} - \boldsymbol{\tau}) = 0, \tag{4.2}$$

$$\frac{\partial \rho(e + u^2/2)}{\partial t} + \nabla \cdot \left\{ \rho\left(h + \frac{u^2}{2}\right)\mathbf{U} - \boldsymbol{\tau} \cdot \mathbf{U} + \mathbf{q} \right\} = 0.$$

Our main concern is the following question: Assuming that a solution of (4.1) and (4.2) has very high gradients inside a "layer," such that when $\varepsilon \to 0$, one gets a shock at a surface (Σ), (Σ) being inside the layer, how does one find the correct asymptotic expansion for ε small for such a solution?

As the case $\varepsilon \neq 0$ is a *singular perturbation* problem of the case $\varepsilon = 0$, it cannot be expected to have an asymptotic expansion uniformly valid everywhere. Accordingly, two types of different asymptotic expansions have to be introduced in order to represent what happens in the "layer" and in the neighborhood of this layer. The first one—the *outer expansion*—will be valid far from (Σ), far with respect to a distance of the order of some power of ε. The second—the *inner expansion*—will be valid inside the layer. Thus we have to obtain these two expansions and *match* them in such a way that they represent the same solution of the Navier–Stokes equations. This kind of technique, introduced by Kaplun and Lagerstrom, is now very well known (Kaplun and Lagerstrom, 1957). Note that we want to obtain the complete expansions to any order. The outer expansion is obtained by starting with fixed geometrical coordinates and making $\varepsilon \to 0$, while the inner expansion requires the introduction of new coordinates based on a normal coordinate counted from (Σ), which is of the order of the dissipation length, say ε. Now there is a second question to be answered, connected with a further use of the results in order to apply them to the solution of some given global problem. If we are interested only in the *outer* expansion, how must the expansions on both sides of (Σ) be related in order to be expansions of the same solution? These relations are the *shock conditions* of the problem.

Classical jump relations (Section II) give the shock conditions to zero order, and classical shock structure (Section III) also gives the inner expansion to zero order. We need to get both to any order.

The usual way of studying such a singular perturbation problem is to proceed by the successive computation of the terms of the two expansions and to write at each step the convenient matching conditions. Here we will use a more direct method. We will *assume* first the analytic properties of the different dependent variables in the neighborhood of (Σ) and show how to define these functions by using the system (4.1), (4.2). The validity of the method rests upon the check at each step of the consistency of the formulas which are found. As it is necessarily the case for every problem for which we have no existence theorem it cannot be rigorously proven that the solution which is found in this way gives the exact correct behavior of the global solution. It says only that it provides a solution which satisfies all the conditions, which contains *no more indeterminacy* than what seems to be required for such a local problem, and which is *fully consistent.* At the moment, no more can be said, but the results can be used safely for any application.

B. Functions NS (Germain and Guiraud, 1962, 1966)

We start with a function $f(z, \varepsilon)$ which *depends on a single coordinate* z and on the (small) positive parameter ε. We look at the case where $\lim_{\varepsilon \to 0} f(z, \varepsilon)$ exists in a domain which contains $z = 0$, except at $z = 0$, where we want to get a jump for the limit. Such a result is obtained for the functions $f(z, \varepsilon)$ which belong to the class NS defined by the following

Definition 4.1. *Let* $f(z, \varepsilon)$ *be defined in the domain* $K(a, \varepsilon_0)$,$[|z| \leq a, 0 \leq \varepsilon \leq \varepsilon_0]$. *The function* $f(z, \varepsilon)$ *is said to belong to the class* NS *if one may write*

$$f(z, \varepsilon) = \tfrac{1}{2}[f^+(z, \varepsilon) + f^-(z, \varepsilon)] + \tfrac{1}{2}[f^+(z, \varepsilon) - f^-(z, \varepsilon)] \operatorname{th} \zeta + \tilde{f}(\zeta, \varepsilon), \tag{4.3}$$

where $z = \varepsilon\zeta$ *and where* f^+ *and* f^- *belong to the class* A *and* $\tilde{f}$ *to the class* B. *these classes being defined below. We will say in such a case that* f^+, f^-, $\tilde{f}$ *define the functions* f *of the class* NS, *if* (4.3) *is valid.*

The presence of $\operatorname{th} \zeta$ is essential as may be seen with the very simple example $f^+ = A$, $f^- = B$, $\tilde{f} = 0$ where A and B are constants; $f(z, \varepsilon)$ tends uniformly towards A for $z \geq \delta > 0$ and towards B for $z \leq -\delta < 0$ however small be the length δ.

Definition 4.2. *A function* $k(z, \varepsilon)$ *belongs to the class* A *if it is defined in* $K(a, \varepsilon_0)$ *and if, in this domain,*

(i) *the following asymptotic expansion is valid*

$$k(z, \varepsilon) = \sum_0^N \varepsilon^n k_n(z) + O(\varepsilon^{N+1}); \tag{4.4}$$

(ii) *one may expand* $k_n(z)$ as

$$k_n(z) = \sum_0^{pn} k_{n,p} z^p + z^{pn+1}\Phi_{pn}(z), \tag{4.5}$$

with $\Phi_{pn}(z)$ *bounded in* $|z| < a$.

DEFINITION 4.3. *A function* $\tilde{f}(\zeta, \varepsilon)$, *defined for* $\varepsilon\zeta \leq a$, $0 \leq \varepsilon \leq \varepsilon_0$ *belongs to the class B if*

(i) *The following asymptotic expansion is valid*

$$\tilde{f}(\zeta, \varepsilon) = \sum_0^N \varepsilon^n \tilde{f}_n(\zeta) + \varepsilon^{N+1}\tilde{\psi}_N(\zeta, \varepsilon). \tag{4.6}$$

(ii) *If* $\exp\{\alpha|\zeta|)\tilde{f}_n(\zeta)\}$ *is defined and uniformly bounded in* $-\infty \leq \zeta \leq +\infty$, *the number* α *and the bound M being independent of n and N.*

(iii) *If* $\exp\{\alpha \operatorname{Inf}(|\zeta|, \varepsilon^{-1}a)\}\tilde{\psi}_N(\zeta, \varepsilon)$ *is also bounded by M.*

In what follows in order to simplify the technical details we assume that N, p_n are infinite, and consequently we will not bother with Φ_{pn} and $\tilde{\psi}_N$. The general study can be found in (Germain and Guiraud, 1966). The following properties are easily proved:

LEMMA 4.1. *If* $k(z, \varepsilon) \in A$, $\tilde{g}(\zeta, \varepsilon) \in B$ *then*

$$k(\varepsilon\zeta, \varepsilon)\tilde{g}(\zeta, \varepsilon) \in B,$$

$$(\operatorname{th} \zeta)k(\varepsilon\zeta, \varepsilon)\tilde{g}(\zeta, \varepsilon) \in B.$$

LEMMA 4.2. *If* $f_1(z, \varepsilon) \in NS$, $f_2(z, \varepsilon) \in NS$ *then*

$$f_1 + f_2 \in NS,$$

$$f_1 f_2 \in NS.$$

LEMMA 4.3. (*Outer Expansion*). *For every* $z \neq 0$, *we can write the following asymptotic expansions*

$$f(z, \varepsilon) \sim \begin{cases} \sum_0^\infty \varepsilon^n f_n^+(z) & \text{if} \quad z > 0, \\ \sum_0^\infty \varepsilon^n f_n^-(z) & \text{if} \quad z < 0. \end{cases} \tag{4.7}$$

Thus, for any $f \in NS$, f_n^+ and f_n^- are uniquely defined.

LEMMA 4.4. (*Inner Expansion*). *For* ζ *fixed, the following asymptotic expansion is valid*

$$f(z, \varepsilon) = f(\varepsilon\zeta, \varepsilon) = \tilde{f}(\zeta, \varepsilon) \sim \sum_0^\infty \varepsilon^n \tilde{f}^{(n)}(\zeta) \tag{4.8}$$

where

$$\bar{f}^{(n)}(\zeta) = \tfrac{1}{2}[Q_n{}^+(\zeta) + Q_n{}^-(\zeta)] + \tfrac{1}{2}[Q_n{}^+(\zeta) - Q_n{}^-(\zeta)]\text{th}\zeta + \tilde{f}_n(\zeta) \quad (4.9)$$

the $Q_n(\zeta)$ being polynomials of order n. More precisely, these polynomials are

$$Q_n{}^\pm(\zeta) = \sum_0^n \zeta^p f_{n-p,\,p}. \quad (4.10)$$

Note that the coefficients $f_{n-p,\,p}$ are defined in (4.5). This result shows that for a given $f \in NS$, the $\tilde{f}_n(\zeta)$ are uniquely defined. It says also precisely what is the behavior for ζ infinite of a term $\tilde{f}_n(\zeta)$ of the inner expansion, i.e. $Q^+(\zeta)$ for $z = +\infty$, $Q^-(\zeta)$ for $\zeta = -\infty$, and how it is related to the behavior for $z = \pm 0$ of the terms of the outer expansion. In other words, it provides the matching conditions to any order.

Definition 4.4. *A function $\tilde{f}(\zeta, \varepsilon)$ defined for $\varepsilon|\zeta| < a$, $0 \leq \varepsilon \leq \varepsilon_0$ and which satisfies* (4.8) *and* (4.9) *will be said to belong to the class C.*

For further use the following notations will be introduced

(i) *Bracket of f*: $[f](\varepsilon) = f^+(0, \varepsilon) - f^-(0, \varepsilon)$.

Let us note immediately that

$$[f](\varepsilon) = \sum_0^\infty \varepsilon^n[f_n] \quad \text{with} \quad [f_n] = f^+_{n,\,0} - f^-_{n,\,0}. \quad (4.12)$$

(ii) *Star of f*:

$$f^*(\varepsilon) = \int_{-\varepsilon^{-1}a}^{\varepsilon^{-1}a} \tilde{f}(\zeta)d\zeta - \frac{1}{2}\int_{-\varepsilon^{-1}a}^{\varepsilon^{-1}a} \frac{d\zeta}{\text{ch}^2\zeta}\int_0^{\varepsilon\zeta} \{f^+(t, \varepsilon) - f^-(t, \varepsilon)\}dt, \quad (4.13)$$

(iii) $$f_\pm{}^*(\varepsilon) = \int_0^{\pm\varepsilon^{-1}a} \tilde{f}(\zeta, \varepsilon)d\zeta - \frac{1}{2}\int_0^{\pm\varepsilon^{-1}a} \frac{d\zeta}{\text{ch}^2\zeta}\int_0^{\varepsilon\zeta} \{f^+(t, \varepsilon) - f^-(t, \varepsilon)\}\, dt, \quad (4.14)$$

(iv) $$F^\pm(z, \varepsilon) = \int_0^z f^\pm(t, \varepsilon)dt + \varepsilon f_\pm{}^*(\varepsilon). \quad (4.15)$$

Then we can prove

Theorem 4.1. *There exists a function $\tilde{F}(\zeta, \varepsilon) \in C$ such that the function*

$$F(z, \varepsilon) = \tfrac{1}{2}[F^+(z, \varepsilon) + F(z, \varepsilon] + \tfrac{1}{2}[F^+(z, \varepsilon) - F^-(z, \varepsilon)]\text{th}\zeta + \varepsilon\tilde{F}(\zeta, \varepsilon) \quad (4.16)$$

belongs to the class NS, and such that its derivative is $f(z, \varepsilon)$. The bracket of F is given by

$$[F](\varepsilon) = \varepsilon f^*(\varepsilon); \quad (4.17)$$

moreover, the star of f, $f^*(\varepsilon) = f_+{}^*(\varepsilon) - f_-{}^*(\varepsilon)$, *admits the following asymptotic expansion*

$$f^*(\varepsilon) \sim \sum_0^\infty \varepsilon^n f_n{}^* \tag{4.18}$$

with

$$f_n{}^* = \int_{-\infty}^{\infty} \tilde{f}_n(\zeta)d\zeta - \frac{1}{2}\int_{-\infty}^{\infty} \frac{d\zeta}{\text{ch}^2\zeta}\int_0^\zeta \{Q_n{}^+(t) - Q_n{}^-(t)\}dt. \tag{4.19}$$

It is easy to check that F^+ and F^- belong to class A. If we compute the z derivative of F given by (4.16), we see that it will be equal to f provided

$$\tilde{F}'(\zeta, \varepsilon) + \frac{F^+(\varepsilon\zeta, \varepsilon) - F^-(\varepsilon\zeta, \varepsilon)}{2\varepsilon\text{ch}^2\zeta} = \tilde{f}(\zeta).$$

Then we can define $\tilde{F}(\zeta, \varepsilon)$ by

$$\tilde{F}(\zeta, \varepsilon) = \int_0^{\alpha(\zeta)} \tilde{f}(\zeta, \varepsilon) - \frac{1}{2\varepsilon}\int_0^{\alpha(\zeta)} \frac{d\zeta'}{\text{ch}^2\zeta'}\int_0^{\varepsilon\zeta'} [f^+(t, \varepsilon) - f^-(t, \varepsilon)]dt$$
$$- \tfrac{1}{2}[f_+{}^*(\varepsilon) + f_-{}^*(\varepsilon)] - \tfrac{1}{2}[f_+{}^*(\varepsilon) - f_-{}^*(\varepsilon)]\text{th}\alpha(\zeta),$$

where

$$\alpha(\zeta) = \begin{cases} \text{Inf}\,(\zeta, a/\varepsilon) & \text{if} \quad \zeta > 0, \\ \text{Sup}\,(\zeta, -a/\varepsilon) & \text{if} \quad \zeta < 0, \end{cases}$$

and it is easy to check that $\tilde{F}$ belongs to the class B. Finally (4.17), (4.18), and (4.19) are direct consequences of (4.15), (4.14), and (4.9) as may be checked with the definitions introduced in Lemmas 4.3 and 4.4.

Now we may state two important consequences for what follows:

Corollary 4.1. *If the two functions* $f(z, \varepsilon)$ *and* $g(z, \varepsilon)$ *belong to the class NS and if they obey the following equation*

$$\int_c^z f(z, \varepsilon)\, dz + g(z, \varepsilon) = \text{const.}, \tag{4.20}$$

then one has $[g](\varepsilon) + \varepsilon f^*(\varepsilon) = 0$.

This condition is the general shock condition for Eq. (4.20).

Corollary 4.2. *If f and g belong to the class NS and obey the relation* (4.20), *then one has*

$$\sum_{p=0}^{n} [g_p] + \sum_{p=0}^{n-1} f_p{}^* = 0. \tag{4.21}$$

This relation is called the "shock condition" up to the order n for Eq. (4.20). The relation $[g_n]+f^*_{n-1}=0$ is the "shock condition" of order n.

Now, we have to show how to use these definitions which are related to functions of a single variable z to answer the questions which have been formulated in the introduction to this section. For that it is necessary to give a few geometrical definitions.

C. Geometrical Remarks

Let us call $(\Sigma(t))$ the surface which carries the discontinuity in perfect fluid flow and which lies inside the "steepening layer" when dissipation effects are present. A point $P \in (\Sigma)$ will be represented as $P(P_0, t)$ where P_0 is its position at some time $t=t_0$. Then the velocity $\mathbf{W}$ of P may be written

$$\mathbf{W}=dP/dt=\mathbf{W}_{\mathrm{t}}+w\mathbf{N}, \tag{4.22}$$

where $\mathbf{N}$ is the unit normal vector to (Σ), $\mathbf{W}_{\mathrm{t}}$ is the tangential component, and $\mathbf{W}$ is independent of the representation which has been chosen in order to describe the motion of P.

At every instant t, let $\mathbf{e}_1$ and $\mathbf{e}_2$ be unit vectors tangent to the lines of curvature of (Σ), x_1 and x_2 the curvilinear coordinates on (Σ) associated with these lines, and z the distance of a point M from (Σ),

$$\mathbf{PM}=z\mathbf{N}.$$

Now

$$\mathbf{I}_{\mathrm{TT}}=\mathbf{e}_1\mathbf{e}_1+\mathbf{e}_2\mathbf{e}_2 \tag{4.23}$$

is the unit tensor on (Σ),

$$(H_1\,dx_1)^2+(H_2\,dx_2)^2$$

is the square of the elementary length on (Σ)

$$\mathbf{K}=\frac{K_1}{H_1}\mathbf{e}_1\mathbf{e}_1+\frac{K_2}{H_2}\mathbf{e}_2\mathbf{e}_2$$

is the tensor of curvature of (Σ), and the mean curvature is $\frac{1}{2}K$ if

$$K=\frac{K_1}{H_1}+\frac{K_2}{H_2}. \tag{4.24}$$

The tensor

$$\mathbf{H}=\mathbf{e}_1\mathbf{e}_1(1+H_1^{-1}K_1z)^{-1}+\mathbf{e}_2\mathbf{e}_2(1+H_2^{-1}K_2z)^{-1} \tag{4.25}$$

is the unit tensor on a surface parallel to (Σ) at a distance z and the operator

$$\boldsymbol{\delta} = \frac{\mathbf{e}_1}{H_1}\frac{\partial}{\partial x_1} + \frac{\mathbf{e}_2}{H_2}\frac{\partial}{\partial x_2} \tag{4.26}$$

is the operator "del" on such a surface. It is easy to check that the spatial "del" is

$$\boldsymbol{\nabla} = \mathbf{N}\frac{\partial}{\partial z} + \mathbf{H}\cdot\boldsymbol{\delta} \tag{4.27}$$

Assume $\partial/\partial t$ is the time derivative in the fixed space and $\delta/\delta t$ is the time derivative in the frame $P(\mathbf{e}_1, \mathbf{e}_2, \mathbf{N})$. If $\boldsymbol{\omega}$ is the rotation of this frame with respect to the fixed space,

$$\partial\mathbf{N}/\partial t = \boldsymbol{\omega} \wedge \mathbf{N}$$

and

$$\frac{\delta}{\delta t} = \frac{\partial}{\partial t} + \mathbf{W}_{\mathrm{T}}\cdot\mathbf{H}\cdot\boldsymbol{\delta} + w\frac{\partial}{\partial t} + z(\boldsymbol{\omega} \wedge \mathbf{N})\cdot\mathbf{H}\cdot\boldsymbol{\delta}, \tag{4.28}$$

because at the instant t, the velocity field of points tied with the frame is

$$\boldsymbol{\Phi} = \mathbf{W} + z(\boldsymbol{\omega} \wedge \mathbf{N}) = \mathbf{W}_{\mathrm{T}} + w\mathbf{N} + z(\boldsymbol{\omega} \wedge \mathbf{N}). \tag{4.29}$$

The preceding formulas are sufficient to understand the results which are presented in the rest of this section. In order to derive them completely more results of differential geometry are needed. The reader who is interested in their derivation may find the details in Germain and Guiraud (1966).

D. Working Assumptions

Let us consider a flow in a region where a shock (Σ) is present when viscosity dissipation is neglected. If we take (small) dissipations into account, we expect expansions like

$$p = p^{(0)} + \varepsilon p^{(1)} + \varepsilon^2 p^{(2)} + \cdots \tag{4.30}$$

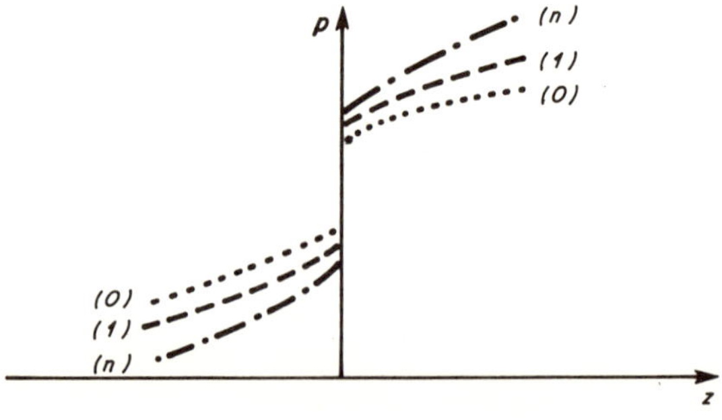

Fig. 11. Outer approximations.

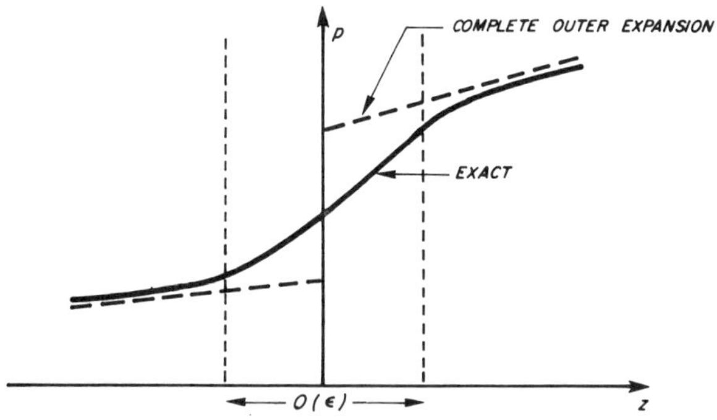

FIG. 12. Comparison of the complete outer expansion and of the exact solution.

to be valid far from (Σ). These expansions, which are the outer expansions, are not valid in the neighborhood of (Σ), but nevertheless they can be mathematically continued up to (Σ); when this is done we obtain results such as those given in Fig. 11, when we keep one, two, or $(n+1)$ terms in the right-hand side of (4.30). The discontinuities which appear for $z=0$ are the jumps we wish to study.

In fact the situation is as in Fig. 12 where the dashed lines are the values of p given by the full outer expansion (4.30) and in solid line is the exact value of p. They differ on a sheet around $z=0$—the "shock layer." In order to compare near $z=0$ the two different behaviors we put $z=\varepsilon\zeta$ in the two functions represented in Fig. 12, expand in ε with ζ fixed (inner expansion), and find the different approximations sketched in Fig. 13 for the outer expansion and in Fig. 14 for the exact solutions. In Fig. 13, on both sides of the shocks we get polynomials of order 0, 1, n. For sufficiently large ζ, i.e. for values of z which are not infinitesimally small, the outer expansion and the exact solution are very close, and we see that the asymptotic behavior of the curves of Fig. 14 are the polynomial curves given in Fig. 13. Indeed these last polynomials are obviously related to the Taylor expansion of the outer solution.

All these heuristic considerations clearly lead to the assumption that

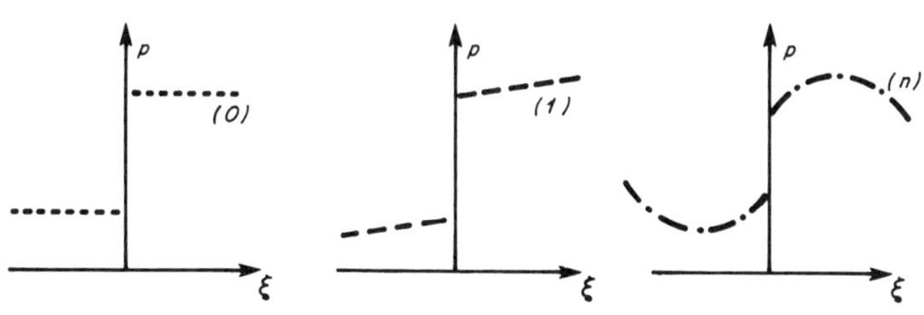

FIG. 13. Inner expansion of the complete outer expansion.

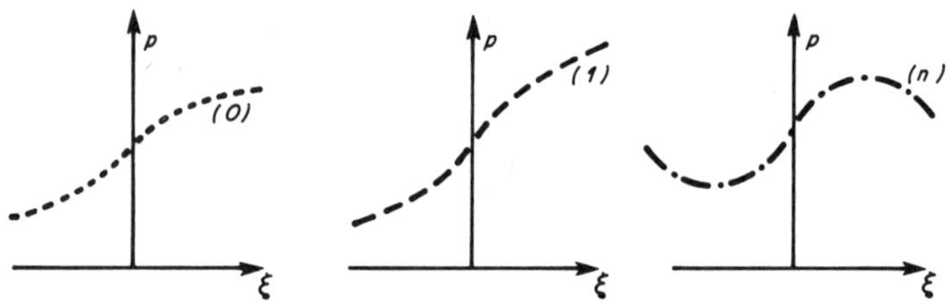

FIG. 14. Inner expansion of the exact solution.

with respect to the z variable, the function p belongs to the class NS. More precisely, we will assume that *every quantity which appears in the description of the flow belongs, as a function of z and ε, to the class NS.* Such a function may depend on the other variables x_1 and x_2 which will be considered as parameters; it may be the result of algebraic operations on the variables p, ρ, $\mathbf{U}$, $\mathbf{W}$, or differentiations such as $\partial/\partial t$, $\partial/\partial z$, $\boldsymbol{\delta}$, or integration on some domain of (Σ). To be more precise, and in order to be able to use the results of Section IV,B we must add to Definitions 4.2 and 4.3 the following complement:

All the properties of these definitions are not altered by any derivation or integration with respect to x_1 and x_2—which are considered as parameters.

Let us note finally that a given quantity of the flow does not have a unique NS representation. Thus we can take another variable $z' = z + \mathscr{L}(\varepsilon)$. $\mathscr{L}(\varepsilon) = a_0 \varepsilon + a_1 \varepsilon^2 + \cdots$, and with this new variable $f(z, \varepsilon) = g(z', \varepsilon)$ will have another NS representation. In a global problem, it is of interest to use this possibility in order to follow the "displacement of the shock" with the dissipation, a phenomenon which arises if at each step the shock is defined, for instance, as the locus of points where the relative velocity of the fluid is sonic. But the determination of $\mathscr{L}(\varepsilon)$ depends on a global problem. Here we consider only the local problem.

E. Study of the Structure

1. *General Equations*

We have to consider the equations (4.2). We note they can be written formally as

$$\frac{\partial \mathscr{A}}{\partial t} + \mathbf{V} \cdot \mathbf{a} = 0, \tag{4.31}$$

where $\mathscr{A}$ is a scalar or a vector and $\mathbf{a}$ a vector or a symmetric tensor respectively. According to (4.28) and (4.27), this equation may be written

$$\frac{\partial}{\partial z}\{\mathbf{a}-\mathscr{A}\mathbf{W})\cdot\mathbf{N}\}+(\mathbf{H}\cdot\boldsymbol{\delta})\cdot\mathbf{a}-(\mathbf{W}_{\mathrm{T}}+z\boldsymbol{\omega}\wedge\mathbf{N})\cdot\mathbf{H}\cdot\boldsymbol{\delta}\mathscr{A}+\frac{\delta\mathscr{A}}{\delta t}=0.$$

Now if we use the inner variable ζ, we can write the inner expansion (see Eq. 4.8) of the left-hand side and obtain

$$\frac{\partial}{\partial\zeta}\{(\overline{\mathbf{a}-\mathscr{A}\mathbf{W}})\cdot\mathbf{N}\}+\varepsilon\left\{\boldsymbol{\delta}\cdot\bar{\mathbf{a}}-\mathbf{W}_{\mathrm{T}}\cdot\boldsymbol{\delta}\overline{\mathscr{A}}+\frac{\delta}{\delta t}\overline{\mathscr{A}}\right\}+O(\varepsilon^2)=0. \quad (4.32)$$

Now, with the notations of formula (4.8), the first term, which will define the *structure of zero order*, is given by

$$(\bar{\mathbf{a}}^{(0)}-\overline{\mathscr{A}}^{(0)}\mathbf{W})\cdot\mathbf{N}=\text{const.} \quad (4.33)$$

The structure of order 1 is defined by the equation

$$\frac{\partial}{\partial\zeta}\{(\overline{\mathbf{a}-\mathscr{A}\mathbf{W}})^{(1)}\cdot\mathbf{N}\}+\boldsymbol{\delta}\cdot\bar{\mathbf{a}}^{(0)}-\mathbf{W}_{\mathrm{T}}\cdot\boldsymbol{\delta}\overline{\mathscr{A}}^{(0)}+\frac{\delta}{\delta t}\overline{\mathscr{A}}^{(0)}=0. \quad (4.34)$$

Obviously, a similar procedure can be applied in order to get the equation which rules the structure of order n.

2. *Shock Structure of Order Zero*

We may apply Eq. (4.33) to each of the equations (4.2) which are all of the form of Eq. (4.31). When writing these equations, one discovers that only the relative velocity $\mathbf{V}=\mathbf{U}-\mathbf{W}$ appears. Let us call $V_{\mathrm{n}}=v$ the normal component of $\mathbf{V}$. Conservation of mass gives

$$\overline{\rho v}^{(0)}=\text{const.}=m. \quad (4.35)$$

The tangential component $\mathbf{V}_{\mathrm{T}}$ obeys the following:

$$\overline{\rho v\mathbf{V}_{\mathrm{T}}}^{(0)}-\mu\,\partial\overline{\mathbf{V}}_{\mathrm{T}}{}^{(0)}/\partial\zeta=\text{const.}$$

But according to (4.9), $\overline{\mathbf{V}}_{\mathrm{T}}{}^{(0)}$ must be bounded in $-\infty<\zeta<+\infty$. It is well known that the only bounded solution of a linear equation like the preceding one is a constant. Then $\overline{\mathbf{V}}_{\mathrm{T}}{}^{(0)}$ is *a constant*. It remains now to write the normal component of the momentum equation and the energy equation. Taking into account the previous results we obtain

$$\begin{aligned}\overline{p+\rho v^2}^{(0)}-(\overline{\lambda+2\mu)\,\partial v/\partial\zeta}^{(0)}&=\text{const.}=P,\\ m(\overline{h+\tfrac{1}{2}v^2})^{(0)}-\bar{\mathbf{k}}^{(0)}\,\partial\overline{T}^{(0)}/\partial\zeta-(\overline{\lambda+2\mu})v\,\partial v^{(0)}/\partial\zeta&=\text{const.}=E.\end{aligned} \quad (4.36)$$

As, according to (4.35), $\bar{v}^{(0)} = m\bar{\tau}^{(0)}$, (4.36) is in fact a system which allows us to define $\bar{\tau}^{(0)}$ and $\bar{T}^{(0)}$ as functions of ζ. Other thermodynamic quantities are computed by the equation of state. As λ and μ are known functions of T it is easy to compute $\bar{\lambda}^{(0)}$ and $\bar{\mu}^{(0)}$ when $\bar{T}^{(0)}$ is known.

The fundamental remark is that this system is completely equivalent to the system which defines the classical shock structure—see, for instance, Eq. (3.40) (with easy changes of notation), and the same conditions must be satisfied, namely that the unknown quantities must be *defined* and *bounded* for every ζ. We have seen that within a translation on ζ the solution exists and is unique. Moreover (see Lemma 4.4) the matching conditions allow us to recover once again the classical shock conditions

$$[p + \rho v^2] = 0, \qquad [h + \tfrac{1}{2}v^2] = 0.$$

3. *Structure of Order* 1

The equation (4.34) may be integrated once to give

$$\overline{(\mathbf{a} - \mathscr{A}\mathbf{W})}^{(1)} \cdot \mathbf{N} = \overline{\mathscr{B}}^{(0)} = -\int_0^\zeta \overline{\mathbf{B}}^{(0)}\, d\zeta + \mathbf{C}, \tag{4.35}$$

where $\overline{\mathscr{B}}^{(0)}$ is a primitive of a quantity $-\overline{\mathbf{B}}^{(0)}$ which depends only on the zero-order shock structure solution and $\mathbf{C}$, a constant. In classical gas dynamics we have five scalar-valued equations like (4.35). That means that the system of equations such as (4.35) which defines the first-order structure *involves five scalar constants* which come from integration.

This system is quite complicated; it is impossible to give here all the calculations which allow us to obtain the result. Let us write

$$(\mathbf{a} - \mathscr{A}\mathbf{W}) \cdot \mathbf{N} = \begin{cases} \gamma, & \text{mass,} \\ \boldsymbol{\mu}, & \text{momentum,} \\ \eta, & \text{energy,} \end{cases}$$

and (4.36)

$$\mathbf{B} = \begin{cases} \mathscr{C}', & \text{mass,} \\ \mathscr{M}', & \text{momentum,} \\ \mathscr{E}', & \text{energy.} \end{cases}$$

In order to exhibit the relative motion we will write the system as

$$\frac{\partial \bar{\gamma}^{(1)}}{\partial \zeta} + \overline{\mathscr{C}}^{(0)} = 0, \qquad \frac{\partial}{\partial \zeta}\overline{(\boldsymbol{\mu} - \gamma \mathbf{W})}^{(1)} + \overline{\mathscr{M}}^{(0)} = 0, \tag{4.37}$$

$$\frac{\partial}{\partial \zeta}\overline{(\eta - \boldsymbol{\mu} \cdot \mathbf{W} + \tfrac{1}{2}\gamma W^2)}^{(1)} + \overline{\mathscr{E}}^{(0)} = 0,$$

with

$$\mathscr{C}' = \mathscr{C}, \quad \mathscr{M}' - \mathscr{C}'\mathbf{W} = \mathscr{M}, \quad \mathscr{E}' - \mathscr{M} \cdot \mathbf{W} + \tfrac{1}{2}\mathscr{C}W^2 = \mathscr{E}. \tag{4.38}$$

Now the full expansions for the first-order structure are obtained with

$$\bar{\gamma}^{(1)} = \overline{\rho v}^{(1)},$$

$$\begin{aligned}\bar{\boldsymbol{\pi}}^{(1)} &= \overline{\rho v \mathbf{V}_{\mathrm{T}}}^{(1)} - \overline{\mu \frac{\partial \mathbf{V}_{\mathrm{T}}}{\partial \zeta}}^{(1)} - \bar{\mu}^{(0)}(\boldsymbol{\delta} \bar{v}^{(0)} - \mathbf{K} \cdot \overline{\mathbf{V}}_{\mathrm{T}}{}^{(0)} + (\boldsymbol{\delta} \mathbf{W}) \cdot \mathbf{N}) \\ &\quad + \mathbf{N}\left\{\overline{p + \rho v^2}^{(1)} - \overline{(\lambda + 2\mu) \frac{\partial v}{\partial \zeta}}^{(1)} - \lambda^{(0)}(K \bar{v}^{(0)} + \boldsymbol{\delta} \cdot \overline{\mathbf{V}}_{\mathrm{T}}{}^{(0)} + \right. \\ &\quad \left. + \boldsymbol{\delta} \cdot \mathbf{W})\right\}, \end{aligned} \tag{4.39}$$

$$\begin{aligned}\bar{\eta}^{(1)} &= \overline{\rho v\left(h + \frac{v^2}{2}\right)}^{(1)} - \overline{k \frac{\partial T}{\partial \zeta}}^{(1)} - \overline{(\lambda + 2\mu) v \frac{\partial v}{\partial \zeta}}^{(1)} - \mu \overline{\mathbf{V}}_{\mathrm{T}}{}^{(0)} \cdot \overline{\frac{\partial \mathbf{V}_{\mathrm{T}}}{\partial \zeta}}^{(1)} \\ &\quad - \bar{\lambda}^{(0)}(K \overline{v^2}^{(0)} + \overline{v\, \boldsymbol{\delta} \cdot \mathbf{V}_{\mathrm{T}}}^{(0)}) - \overline{\mu(\mathbf{V} \cdot \boldsymbol{\delta} v - \mathbf{V}_{\mathrm{T}} \cdot \mathbf{K} \cdot \mathbf{V}_{\mathrm{T}})}^{(0)} \\ &\quad - \overline{\lambda v}^{(0)} \boldsymbol{\delta} \cdot \mathbf{W} - \overline{\mu \mathbf{V}_{\mathrm{T}}}^{(0)} \cdot \boldsymbol{\delta}\, \mathbf{W} \cdot \mathbf{N}, \end{aligned}$$

and

$$\begin{aligned}\bar{\mathscr{C}}^{(0)} &= \boldsymbol{\delta} \cdot \overline{\rho \mathbf{V}_{\mathrm{T}}}^{(0)} + K \overline{\rho v}^{(0)} + \overline{\frac{\delta \rho}{\delta t}}^{(0)} + \bar{\rho}^{(0)}\, \boldsymbol{\delta} \cdot \mathbf{W}, \\ \bar{\mathscr{M}}^{(0)} &= \boldsymbol{\delta} \cdot \overline{(\rho \mathbf{V}\mathbf{V})}^{(0)} + \boldsymbol{\delta}\overline{\left(p - \lambda \frac{\partial v}{\partial \zeta}\right)}^{(0)} - 2K \overline{\mu \frac{\partial v}{\partial \zeta}}^{(0)} \mathbf{N} \\ &\quad + \overline{\frac{\delta \rho \mathbf{W}}{\delta t}}^{(0)} + \overline{\frac{\delta \rho \mathbf{V}}{\delta t}}^{(0)} + \overline{\rho \mathbf{V}}^{(0)} \boldsymbol{\delta} \cdot \mathbf{W} + \overline{\rho \mathbf{V}}^{(0)} \cdot \boldsymbol{\delta} \mathbf{W} \\ \bar{\mathscr{E}}^{(0)} &= \boldsymbol{\delta} \cdot \left\{\overline{\left[\rho(h + \tfrac{1}{2} v^2) - \lambda \frac{\partial v}{\partial \zeta}\right] \mathbf{V}}^{(0)}\right\} - K\left\{\overline{2\mu v \frac{\partial v}{\partial \zeta} + k \frac{\partial T}{\partial \zeta}}^{(0)}\right\} \\ &\quad + \frac{\delta}{\delta t} \overline{\rho(e + \tfrac{1}{2} v^2)}^{(0)} + \overline{\rho \mathbf{V}}^{(0)} \frac{\delta \mathbf{W}}{\delta t} + \left\{\overline{\rho(h + \tfrac{1}{2} v^2) - \lambda \frac{\partial v}{\partial \zeta}}^{(0)}\right\} \boldsymbol{\delta} \cdot \mathbf{W} \\ &\quad + \overline{\rho \mathbf{V}}^{(0)} \cdot \boldsymbol{\delta} \mathbf{W} \cdot \overline{\mathbf{V}}^{(0)}. \end{aligned} \tag{4.40}$$

Note that this system is linear with respect to the set of quantities $\bar{\rho}^{(1)}$, $\bar{v}^{(1)}$, etc. since

$$\overline{f(a) bc}^{(1)} = f(\bar{a}^{(0)}) \bar{b}^{(0)} \bar{c}^{(1)} + f(\bar{a}^{(0)}) \bar{b}^{(1)} \bar{c}^{(0)} + f'(\bar{a}^{(0)}) \bar{a}^{(1)} \bar{b}^{(0)} \bar{c}^{(0)}$$

Then if we define a vector $\overline{\mathbf{X}}^{(1)}$ as the collection of quantities

$$\overline{\mathbf{X}}^{(1)} = \{\bar{\rho}^{(1)}, \bar{p}^{(1)}, \bar{T}^{(1)}, \bar{\lambda}^{(1)}, \bar{\mu}^{(1)}, \bar{k}^{(1)}, \bar{v}^{(1)}, \overline{\mathbf{V}}_{\mathrm{T}}^{(1)}\},$$

the system which defines the structure of order 1 is of the form

$$A(\overline{\mathbf{X}}^{(0)}) \overline{\mathbf{X}}^{(1)} + B(\overline{\mathbf{X}}^{(0)})\, d\overline{\mathbf{X}}^{(1)}/d\zeta = \overline{\mathbf{Y}}^{(1)}(\overline{\mathbf{X}}^{(0)}).$$

4. *General Properties of the Structure of Order n*

Now it is quite clear that the system which rules the nth-order structure is of a similar form, namely

$$A(\overline{\mathbf{X}}^{(0)})\overline{\mathbf{X}}^{(n)} + B(\overline{\mathbf{X}}^{(0)})\, d\overline{\mathbf{X}}^{(1)}/d\zeta = \overline{\mathbf{Y}}^{(n)}(\overline{\mathbf{X}}^{(0)}, \overline{\mathbf{X}}^{(1)}, \ldots, \overline{\mathbf{X}}^{(n-1)}). \quad (4.41)$$

It is also easy to show more precisely that $\overline{\mathbf{Y}}^{(n)}$ is of the form

$$\overline{\mathbf{Y}}^{(n)} = \overline{\mathbf{F}}^{(n)}(\overline{\mathbf{X}}^{(0)}, \overline{\mathbf{X}}^{(1)}, \ldots, \overline{\mathbf{X}}^{(n-1)}) + \int^{\zeta} \overline{\mathbf{G}}^{(n-1)}(\overline{\mathbf{X}}^{(0)}, \overline{\mathbf{X}}^{(1)}, \ldots, \overline{\mathbf{X}}^{(n-1)})\, d\zeta, \quad (4.42)$$

where $\overline{\mathbf{F}}^{(n)}$ and $\overline{\mathbf{G}}^{(n-1)}$ are functionals of $\overline{\mathbf{X}}^{(0)}, \ldots, \overline{\mathbf{X}}^{(n-1)}$ which involve algebraic operations, derivatives with respect to t and ζ, and derivatives with respect to x_1 and x_2 by means of the operator $\boldsymbol{\delta}$. Note that at each step $n (n > 1)$, *five new constants* are introduced in the equations written in (4.2). In what follows, we assume that the choice of these constants has been made.

We arrive now at the critical point of the study. All that has been done until now rests on the assumption that every quantity arising in the theory belongs to the class NS with respect to the variable ζ, the other variables being considered merely as parameters. With this assumption it has been shown how to compute each term of the inner asymptotic expansion. Then we have to check that each of these terms has in fact the properties required by the different terms of the inner expansions, i.e., the properties which are stated in Lemma 4.4 (see Eqs. 4.8, 4.9, and 4.10). If we remark in particular that the matching condition (4.10) allows one to define the coefficients (for small z) of the outer expansions, it is clear that if we can prove that the $\bar{f}^{(n)}(\zeta)$ are of the form given by (4.9), then at least formally $\sum_0^m \varepsilon^n \bar{f}^{(n)}(\zeta)$ is the inner asymptotic expansion of a function NS. The consistency of the theory is then given by the following

THEOREM 4.2. *Any solution* $\overline{\mathbf{X}}^{(n)}$ *of* (4.41) *which for* $\zeta = +\infty$ *and* $\zeta = -\infty$ *is bounded by some polynomial can be written*

$$\overline{\mathbf{X}}^{(n)} = \tfrac{1}{2}(\overline{\mathbf{X}}_+^{(n)} + \overline{\mathbf{X}}_-^{(n)}) + \tfrac{1}{2}(\overline{\mathbf{X}}_+^{(n)} - \overline{\mathbf{X}}_-^{(n)}) \operatorname{th} \zeta + \tilde{\mathbf{X}}^{(n)}(\zeta),$$

where $\overline{\mathbf{X}}_\pm^{(n)}(\zeta)$ *are vector-valued polynomials in* ζ, *of degree n at most, and* $\tilde{\mathbf{X}}(\zeta)$ *is a function which at* $\zeta = +\infty$ *or* $-\infty$ *decreases as an exponential* (*of negative power*). *It is also possible to write in a similar way*

$$\overline{\mathbf{F}}^{(n)} = \tfrac{1}{2}(\overline{\mathbf{F}}_+^{(n)} + \overline{\mathbf{F}}_-^{(n)}) + \tfrac{1}{2}(\overline{\mathbf{F}}_+^{(n)} - \overline{\mathbf{F}}_-^{(n)}) \operatorname{th} \zeta + \tilde{\mathbf{F}}^{(n)},$$
$$\overline{\mathbf{G}}^{(n)} = \tfrac{1}{2}(\overline{\mathbf{G}}_+^{(n)} + \overline{\mathbf{G}}_-^{(n)}) + \tfrac{1}{2}(\overline{\mathbf{G}}_+^{(n)} - \overline{\mathbf{G}}_-^{(n)}) \operatorname{th} \zeta + \tilde{\mathbf{G}}^{(n)},$$
$$\overline{\mathbf{Y}}^{(n)} = \tfrac{1}{2}(\overline{\mathbf{Y}}_+^{(n)} + \overline{\mathbf{Y}}_-^{(n)}) + \tfrac{1}{2}(\overline{\mathbf{Y}}_+^{(n)} - \overline{\mathbf{Y}}_-^{(n)}) \operatorname{th} \zeta + \tilde{\mathbf{Y}}^{(n)},$$

where n is again an upper bound of the degree of the polynomials $\overline{\mathbf{F}}_{\pm}^{(n)}$, $\overline{\mathbf{G}}_{\pm}^{(n)}$, $\overline{\mathbf{Y}}_{\pm}^{(n)}$. *Moreover*

$$[\overline{\mathbf{Y}}^{(n)}] = [\overline{\mathbf{F}}^{(n)}] + \overline{\mathbf{G}}^{(n-1)*}.$$

Finally, when the five constants arising have been chosen, the only indeterminacy which remains in $\overline{\mathbf{X}}^{(n)}$ *is in* $\tilde{\mathbf{X}}^{(n)}(\zeta)$, $\tilde{\mathbf{X}}^{(n)}(\zeta)$ *is defined up to an (additive) vector of the form* $\alpha\hat{\mathbf{X}}(\zeta)$ *where* $\hat{\mathbf{X}}(\zeta)$ *is the unique solution of the homogeneous system:*

$$A\hat{\mathbf{X}} + B\, d\hat{\mathbf{X}}/d\zeta = 0,$$

which is zero for $\zeta = -\infty$ *and* $\zeta = +\infty$.

The proof is not too difficult and goes by induction on n (Germain and Guiraud, 1966). The essential ideas are the following: For a given n, $\overline{\mathbf{X}}_{+}^{(n)}$ and $\overline{\mathbf{X}}_{-}^{(n)}$ are defined by systems like (4.41) in which one puts $\zeta = +\infty$ and $\zeta = -\infty$. Then it may be shown that $\tilde{\mathbf{X}}^{(n)}$ is the solution of a system like

$$A\tilde{\mathbf{X}} + B\, d\tilde{\mathbf{X}}/d\zeta = \tilde{\mathbf{Z}},$$

where $\tilde{\mathbf{Z}}$ is exponentially small at infinity. Again the behavior at infinity is fundamental. In the proof, one has to use the solutions of the system which defines the classical shock structure which has been studied in Section III. Let us repeat that this theorem shows the consistency of the theory and in particular the validity of the working assumptions.

F. Shock Conditions

1. *Matching Method*

When the structure of the shock is known up to the order n, it is easy to obtain the jump relations up to the same order. The shock conditions to the zero order are, as already stated, the Rankine–Hugoniot conditions. Now the shock conditions of order n may be obtained by applying Corollary 4.2 of Theorem 4.1 to (4.41). For instance, for $n = 1$ we get

$$\left\{A\overline{\mathbf{X}}_{+}^{(1)} + B\frac{d\overline{\mathbf{X}}_{+}^{(1)}}{d\zeta}\overline{\mathbf{F}}_{+}^{(1)}\right\}_{\zeta=+\infty} - \left\{A\overline{\mathbf{X}}_{-}^{(1)} + B\frac{d\overline{\mathbf{X}}_{-}^{(1)}}{d\zeta} - \overline{\mathbf{F}}_{-}^{(1)}\right\}_{\zeta=-\infty} - \mathbf{G}^{(0)*} = 0. \tag{4.43}$$

Taking account of the results given for this order—in particular (4.39) and (4.40)—it is not difficult to write these conditions explicitly. The same, in principle, can be done for any n. Let us emphasize that for the time being (4.43) shows that *in order to write the shock conditions up to the order* 1, *it is necessary to know the solution of the structure of order* 0 (classical shock structure).

This result which is extended to any order is a direct consequence of (4.17).

2. *Direct Method*

Before writing the explicit shock conditions up to the order 1, let us show how the shock conditions may be obtained (up to any order)—and in a slightly easier way—directly by a generalization of the method used in Section II. We will use the domain $(\mathscr{D}_\delta)$, bounded by the union (S) of (S_δ), $(S_{-\delta})$, (S_b) (see Fig. 6), but it is convenient here to take this domain fixed in the frame tied with (Σ); $P(\mathbf{e}_1, \mathbf{e}_2, \mathbf{N})$ introduced in Section IV, C. Then the conservation laws take the following form (see formula 2.2)

$$\frac{\delta}{\delta t}\iiint_{(\mathscr{D})} \mathscr{A}\, dv + \iint_{(S)} \mathbf{b}\cdot\mathbf{n}\, d\sigma = 0, \tag{4.44}$$

with

$$\mathbf{b} = \mathbf{a} + \mathscr{A}(\mathbf{U} - \mathbf{\Phi}),$$

where $\mathbf{a}$ and $\mathscr{A}$ are the same as in (4.31) and $\mathbf{\Phi}$ is the velocity of a point tied with the frame given by (4.29).

Using the geometrical definitions of Section IV, C (4.44) when applied to $(\mathscr{D}_\delta)$ gives

$$\left\{\iint_{\Sigma_0} \mathbf{b}\cdot\mathbf{N}(1+zK)\, d\sigma\right\}_{z=-\delta}^{z=+\delta} + \int_{-\delta}^{\delta} dz\left\{\int_\Gamma \mathbf{b}\cdot\mathbf{\Lambda}\cdot\mathbf{v}\, ds + \frac{\delta}{\delta t}\iint_{\Sigma_0} \mathscr{A}(1+zK)\, d\sigma\right\} = 0, \tag{4.45}$$

where (Γ) is the boundary of (Σ_0)—the intersection of (Σ) and $(\mathscr{D}_\delta)$, $\mathbf{v}$ is the unit vector normal to (Γ)—directed outside $(\mathscr{D}_\delta)$—and tangent to (Σ), and $\mathbf{\Lambda}$ is an operator which may be written formally as

$$\mathbf{N}\times(\mathbf{I} + z\mathbf{K})\times\mathbf{N}.$$

For a *fixed* (Σ_0), (4.45) is of the form

$$F(z_2; \mathscr{A}_0, t) - F(z_1; \mathscr{A}_0, t) + \int_{z_1}^{z_2} f(z, \mathscr{A}_0, t)\, dz = 0,$$

and by Theorem 4.1, the general condition may be written

$$[F](\varepsilon) + \varepsilon f^* = 0.$$

It remains to perform the calculations. First, it can be shown that it is possible to commute the operations "Bracket" and "star" with the integral sign in a fixed domain. Thus we get

$$\iint_{\Sigma_0} [\mathbf{b} \cdot \mathbf{N}]\, d\sigma + \int_{\Gamma} (\mathbf{b} \cdot \mathbf{\Lambda})^* \cdot \mathbf{\nu}\, ds + \frac{\delta}{\delta t} \iint_{\Sigma_0} (\mathscr{A}(1 + z\mathrm{K}))^*\, d\sigma = 0.$$

Finally, we want to put all the left-hand side in the form of a *single* integral over Σ_0

$$\iint_{\Sigma_0} \mathbf{C}\, d\sigma = 0$$

in such a way that, by application of the fundamental lemma of continuous mechanics the shock conditions are directly obtained as $\mathbf{C} = 0$. Finally, one gets

$$[\mathbf{b} \cdot \mathbf{N}] + \varepsilon \left\{ \boldsymbol{\delta} \cdot \mathbf{b}^{\mathrm{T}*} - K\mathbf{N} \cdot \mathbf{b} + \frac{\delta \mathscr{A}^*}{\delta t} + \mathscr{A}^* \boldsymbol{\delta} \cdot \mathbf{W} \right\} = 0 \qquad (4.46)$$

for the shock condition up to the order 1 associated with the conservation law written as in (4.44). The shock condition of order 0 is $[\mathbf{b} \cdot \mathbf{N}] = 0$, which is the same result as in Theorem 2.1. Of course a similar method works for any n.

3. *Explicit Formulas in Gas Dynamics*

The shock conditions up to order 1 are

$$[\rho v] + \varepsilon \left\{ \boldsymbol{\delta} \cdot (\rho^* \mathbf{V}_{\mathrm{T}}) + \rho^* \boldsymbol{\delta} \cdot \mathbf{W} + \frac{\delta \rho^*}{\delta t} \right\} = 0,$$

$$\begin{aligned}
[p\mathbf{N} + \rho v \mathbf{V}] + \varepsilon \Big\{ & \boldsymbol{\delta} \cdot \rho^* \mathbf{V}_{\mathrm{T}} \mathbf{V}_{\mathrm{T}} + \boldsymbol{\delta} \left(p - \lambda \frac{\partial v}{\partial \zeta} \right)^* - K \left(p - \lambda \frac{\partial v}{\partial \zeta} \right)^* \mathbf{N} \\
& - \left[\lambda (\boldsymbol{\nabla} \cdot \mathbf{V}) \mathbf{N} + \mu \frac{\partial \mathbf{V}}{\partial \zeta} + \mu (\boldsymbol{\nabla} \mathbf{V}) \cdot \mathbf{N} \right] \\
& + \rho^* \mathbf{V}_{\mathrm{T}}\, \boldsymbol{\nabla} \cdot \mathbf{W} + \rho^* \mathbf{V}_{\mathrm{T}} \cdot \boldsymbol{\nabla} \mathbf{W} + \rho^* \frac{\delta \mathbf{W}}{\delta t} + \frac{\delta \rho \mathbf{V}_{\mathrm{T}}^*}{\delta t} \\
& - [\lambda (\boldsymbol{\nabla} \cdot \mathbf{W}) \mathbf{N} + \mu (\boldsymbol{\nabla} \mathbf{W}) \cdot \mathbf{N}] \Big\} = 0,
\end{aligned}$$

$$[\rho v(h + \tfrac{1}{2}v^2)] + \varepsilon\left\{\boldsymbol{\delta} \cdot \left[\left(\rho\left(h + \frac{v^2}{2}\right) - \lambda \frac{\partial v}{\partial \zeta}\right)^* \mathbf{V}_{\mathrm{T}}\right]\right.$$

$$- \left[\lambda v \boldsymbol{\nabla} \cdot \mathbf{V} + \mu \mathbf{V} \cdot \frac{\partial \mathbf{V}}{\partial z} + \mu \mathbf{V} \cdot (\boldsymbol{\nabla}\mathbf{V}) \cdot \mathbf{N}\right]$$

$$+ \rho^* \mathbf{V}_{\mathrm{T}} \cdot (\boldsymbol{\nabla}\mathbf{W}) \cdot \mathbf{V}_{\mathrm{T}} + \left[\rho(h + \tfrac{1}{2}v^2) - \lambda \frac{dv}{d\zeta}\right]^* \boldsymbol{\nabla} \cdot \mathbf{W}$$

$$+ \rho^* \mathbf{V}_{\mathrm{T}} \cdot \frac{\delta \mathbf{W}}{\partial t} + \frac{\delta}{\delta t}[\rho(e + \tfrac{1}{2}v^2)]^*$$

$$\left. - [\lambda v \boldsymbol{\nabla} \cdot \mathbf{W} + \mu \mathbf{V} \cdot \boldsymbol{\nabla}\mathbf{W} \cdot \mathbf{N}]\right\} = 0.$$

These formulas are very complicated. A simplification is obtained when the flow is steady: all the terms which contain **W** or $\delta/\delta t$ are then absent. Note in particular the Galilean invariance that if a constant vector is added to **W** these formulas do not change.

Similar formulas have been obtained by Sedov *et al.* (1953), though incompletely, because they have not taken into account the "star terms"; in other words, they forget the influence of the thickness of the shock. This defect was noticed by Probstein and Kemp (1963), but they do not give the exact expressions for the thickness effect. For the stationary case the correct expressions have been published for the first time independently by Chow and Ting (1961) and by Germain and Guiraud (1960).

Let us emphasize once more that the "star terms" depend on the complete zero-order structure, through formulas which have some similarity to the formulas which give for instance the momentum thickness in boundary layer theory.

G. Final Remarks

In this section we have attempted to describe the complete asymptotic behavior of the flow in the neighborhood of a shock wave. Only local properties have been studied. The consistency of the scheme has been shown. We have noted all the indeterminacies which are met in this treatment, but we have not tried to discuss their significance. It is clear for instance that if we make a translation in the shock structure of order zero, we change the value of the "star terms." But as was said at the end of Section IV, D, the determination of the position of the shock at each order is something which has to be done when one solves a global problem. It is believed that the indeterminacy which has been found corresponds exactly to the one which is required by the necessity to obtain a variation in the position of a shock in a global problem. Such a conjecture is true, as was shown by Oberai (1964), who has applied these formulas for computing the hypersonic flow around a sphere in a slightly dissipative medium.

Of course the methods which have been described may be applied to other dissipative phenomena. For instance, in Peyret (1963) the shock conditions have been written up to the first order in classical MFD.

REFERENCES

ANDERSON, J. E. (1963). "Magnetohydrodynamic Shock Waves." MIT Press, Cambridge, Massachusetts.

BAZER, J., and ERICKSON, W. B. (1959). Hydromagnetic shocks. *Astrophys. J.* **129**, 758–785.

BURGERS, J. M. (1948). A mathematical model illustrating the theory of turbulence. *Advanc. Appl. Mech.* **1**, 171–179.

BURGERS, J. M. (1957). Perturbation of a shock wave into a magnetic field. *In* "Magnetohydrodynamics" (K. M. Landshoff, ed.), pp. 36–56. Stanford Univ. Press, Stanford, California.

CHOW, R., and TING, L. (1961). Higher order theory of curved shock. *J. Aerosp. Sci.* 428–430.

COLE, J. D. (1951). On a quasilinear parabolic equation occurring in aerodynamics. *Quart. Appl. Math.* **9**, 226–236.

COWLING, T. G. (1957). "Magnetohydrodynamics." Wiley (Interscience), New York.

GERMAIN, P. (1959). Contribution à la théorie des ondes de choc en magnétodynamique des fluides. *Publ. O.N.E.R.A.* No. 97.

GERMAIN, P. (1960a). L'équation de Burgers et ses applications à la théorie des ondes de choc. *Cah. Phys.* **119**, 285–299.

GERMAIN, P. (1960b). Shock waves and shock wave structure in magnetofluid dynamics. *Rev. Mod. Phys.* **32**, 951–958.

GERMAIN, P. (1967). A model of some plasma shock structures. *Magneto-Fluid and Plasma Dynamics, Proc. Symp. Appl. Math.*, **18**, 1967 pp. 17–45.

GERMAIN, P., and BADER, R. (1953). Unicité des éncoulements avec chocs dans la mécanique de Burgers. *Jahrb. Wiss. Gese. Luftfahrt* pp. 144–148.

GERMAIN, P., and GUIRAUD, J. P. (1960). Conditions de choc dans un fluide doué de coefficients de viscosité et de conductibilité thermique faibles mais non nuls. *Co.Re. Acad. Sci.* **250**, 1965–1967.

GERMAIN, P., and GUIRAUD, J. P. (1962). Conditions de choc et structure des ondes de choc dans un ecoulement stationnaire de fluide dissipatif. *Publ. O.N.E.R.A.* No. 105.

GERMAIN, P., and GUIRAUD, J. P. (1966). Conditions de choc et structure des ondes de choc dans un écoulement non stationnaire de fluide dissipatif. *J. Math. Pures Appl.* **45**, 311–358.

HAYES, W. D. (1958). The basic theory of gas-dynamics discontinuities. "Fundamental of Gas Dynamics," pp. 416–481. Princeton Univ. Press, Princeton, New Jersey.

HAYES, W. D. (1960). "Gas Dynamics Discontinuities." Princeton Univ. Press, Princeton, New Jersey (see, also, Hayes, 1958).

HOPF, E. (1950). The partial differential equation $u_t + uu_x = \mu u_{xx}$. *Commun. Pure Appl. Math.* **3**, 201–230.

Kaplun, S., and Lagerstrom, P. A. (1957). Asymptotic expansions of Navier–Stokes solutions for small Reynolds numbers. *J. Rat. Mech. Anal.* **6**, 585–593.

Kulikovskii, A. G. (1962a). On the structure of shock waves in magnetohydrodynamics with arbitrary dissipation law. *Prikl. Mat. Mekh.* **26**, 273–279; see *J. Appl. Math. Mech.* (*USSR*) **26**, 392–400 (1962).

Kulikovskii, A. G. (1962b). The structure of shock waves. *Prikl. Mat. Mekh.* **26**, 631–641; see *J. Appl. Math. Mech.* (*USSR*) **26**, 950–964 (1962).

Kulikovskii, A. G., and Liubimov, G. A. (1962). The structure of a magnetohydrodynamic shock wave in a gas with an anisotropic conductivity. *Prikl. Mat. Mekh.* **26**, 791–792; see *J. Appl. Math. Mech.* (*USSR*) **26**, 1197–1199 (1962).

Ladyzhenskaya, O. A. (1956). *Dokl. Akad. Nauk SSSR* **111**, 291–294.

Lax, P. D. (1957). Hyperbolic systems of conservation laws. *Commun. Pure Appl. Math.* **10**, 537–566.

Liepmann, H. W., and Roshko, A. (1957). "Elements of Gas-Dynamics." Wiley, New York.

Lighthill, M. J. (1956). Viscosity effects in sound waves of finite amplitude. *In* "Surveys in Mechanics. Taylor Anniversary Volume" (G. K. Batchelor and R. M. Davies, eds.), pp. 250–351. Cambridge Univ. Press, London and New York.

Ludford, G. S. S. (1959). The structure of a hydromagnetic shock in steady plane motion. *J. Fluid Mech.* **5**, 67–80.

Marshall, W. (1955). The structure of magnetohydrodynamic shock wave. *Proc. Roy. Soc., Ser. A* **233**, 367–376.

Morton, K. W. (1962). Large amplitude compression waves in adiabatic two-fluid model of a collision-free plasma. *J. Fluid Mech.* **14**, 369–384.

Morton, K. W. (1964). Finite amplitude compression waves in a collision-free plasma. *Phys. Fluids* **7**, 1800–1815.

Oberai, M. N. (1964). Ecoulement hypersonique visqueux autour d'un obstacle de révolution à nez arrondi, en atmosphère modérément raréfiée. *J. Mec.* **3**, 173–214.

Oleinik, O. A. (1956). *Dokl. Akad. Nauk SSSR* **109**, 1098–1101.

Peyret, R. (1963). Conditions de choc pour un écoulement de magnétodynamique des fluides faiblement dissipatifs. *C. R. Acad. Sci.* **256**, 4156–4158.

Probstein, R. F., and Kemp, N. H. (1963). Viscous aerodynamic characteristics in a hypersonic rarefied gas glow. *J. Aerosp. Sci.* **27**, 174–193.

Sedov, L. J., Mikhailova, M. P., and Chernyi, G. G. (1953). *Vestn. Mosk. Univ., Ser. Fiz.-Mat. Estest. Nauk* **8**, No. 3, 95–100.

Shercliff, J. A. (1960). One-dimensional magnetogasdynamics in oblique fields. *J. Fluid Mech.* **9**, 481–565.

Weyl, H. (1949). Shock waves in arbitrary fluids. *Commun. Pure Appl. Math.* **2**, 103–122.

Interplanetary Gas Dynamics

V. C. LIU

The University of Michigan, Ann Arbor, Michigan

I. Introduction

The study of interaction between the earth's magnetic field and a solar particle stream in the interplanetary space dated back at least four decades when Chapman and Ferraro (1931) initiated their simplified plasma interaction model of the geomagnetic storm problem. The interplanetary space and also the outer region of the geomagnetic field were then considered void of any gas until the advent of the solar corpuscular stream, an electrically neutral beam of protons and electrons. The existence of a continual outflow of solar plasma, or the solar wind, was inferred by Biermann (1951) in his cometary studies and the concept of solar wind was formally advanced by Parker (1958). These advances in astrophysics laid the foundation for studies of the dynamics of gases under interplanetary conditions to gain further physical insight into related geo- and astrophysical phenomena of interest. This special branch of the physics of fluids is designated herein the interplanetary gas dynamics, the domain of which encompasses various gas-dynamical consequences pertaining to solar-terrestrial relations. On the solar side, it includes the genesis and development of the solar wind and other solar plasma emissions; on the terrestrial end it concludes, among others, the dynamics of the Van Allen radiation belt and the auroral displays. The term "gas dynamics" which is used here in a very broad sense includes plasma kinetics; furthermore, the word "fluid" and "gas" are often used interchangeably.

The interplanetary gas is a high temperature plasma in motion as a result of hydrodynamic expansion of the solar atmosphere accompanied by a prodigious outpouring of solar radiation in a very broad spectrum. This plasma medium is further complicated by the presence of an interplanetary magnetic field which is also primarily of solar origin. When a magneto-plasma stream meets an obstacle (a planet, comet, spacecraft, or a meteorite) special fluid-dynamical interaction develops that could cause the display of a cosmic phenomenon. A measure of this gas-dynamical response has often been used to infer the physical state of the interplanetary medium on the basis of gas-dynamical analysis. That in the past many astrophysical studies tended toward qualitative description of the phenomena of interest is partly due to the lack of precise quantitative observations. The advent of space exploration, which makes direct measurements possible, and the advance of radio astronomy, which studies an extraterrestrial object by measuring its radio emission, stimulate much needed theoretical efforts to serve as guidance to experimental interplanetary studies. Although the present study is restricted essentially to the theoretical aspect of the subject matter, one should never lose sight of the dictum that ultimately all of the theoretical results of significance must be related to the observations in one way or another. This has a twofold implication: the theoretician must keep the observational results in mind, and there is a continuing need for new and relevant observations as the outcome of the theoretical understanding of the phenomenon in question.

Admittedly gas dynamics is only one of the many aspects of an astrophysical phenomenon. However, in view of the fact that the universe is abundantly filled with gases, the consequence of any disturbances, created either by a moving solid body or by a propagating wave pulse in the medium, is most likely involved somehow with gas dynamics in its broad interpretation. It is found that the methods of generalized fluid mechanics, which include gas kinetics and continuum hydrodynamics as parts of its discipline, have been proved extremely useful in the solution of some contemporary astrophysical problems. Some contributions resulting from fluid-mechanical considerations can be briefly cited here. The hydrodynamic theory of solar wind (Parker, 1958) is an obvious case in point. It extends the frontier of fluid-mechanical research into the domain of astrophysical enquiry through the study of the dynamics of gases under cosmic conditions, particularly in the interplanetary space. The problem of bow shock, which is a gas-dynamical discontinuity of the solar wind in front of a planet (Gold, 1955), has been treated effectively by the theory of fluid mechanics (Axford, 1962; Kellogg, 1962) before the shock was observed. The understanding of whistler* propagation in the ionosphere has been enhanced considerably by the hydromagnetic considerations of wave dispersion and dissipation in a magnetofluid (Ginzburg, 1964). Plasma dynamics plays an indispensible role in the design and evaluation of the *in situ* measurements of the ionosphere (Liu, 1969).

The scope of the present study is limited to the discussion of the fundamental aspects which are essential to the apprehension of the fluid-mechanical basis of some interplanetary phenomena and to the physical concepts which are unique under cosmic conditions. Attention is drawn to those aspects of interplanetary gas dynamics which have close analogy in classical gas dynamics. After a brief discussion of some essential observational results on the relevant physical properties of the interplanetary medium (as presented in this section), a review of the methods of fluid mechanics is presented which covers the treatments of both the extremely rarefied and the continuous media (Section II). A paradoxical characteristic of the interplanetary gas, namely its quasi-continuum behavior in spite of the rarity of collisions, is studied and interpreted in view of the collective behavior of the charged particles (Section III). This is of basic importance to the hydrodynamic theory of the interplanetary gas which, near the earth's orbit, has collisional mean free path of the order of 1 AU (astronomical unit†). The hydrodynamic expansion of the solar corona is discussed and its analogy to the Laval nozzle flow is shown in Section IV. A wave mechanism of mechanical energy transport and dissipation which apparently serves as the primary cause of corona heating is introduced in Section V.

* A special electromagnetic wave mode usually caused by lightning strokes which travel along the geomagnetic field into outer space and back to the other hemisphere through the ionosphere.

† 1 AU $= 1.5 \times 10^{11}$ m.

This special mode of energy transfer is a unique trait of a cosmic medium. In the last two sections (VI and VII), free expansion of neutral particle clouds and their interplanetary applications are presented.

Though the external magnetic field is an essential aspect of interplanetary gas dynamics, the internal electrostatic field is still considered as the fundamental field, the effect of which is augmented or reduced by any externally applied field which may be present. In order to limit the present review to a proper extent, it is decided to exclude the effect of the interplanetary magnetic field from the present discussion. This somewhat influences the selection of topics to be included herein. For instance the dynamics of magnetosphere, over which the magnetic field has a dominant influence, is not included. This exclusion seems justifiable in view of the existence of some recent excellent reviews on the dynamics of magnetosphere (Parker, 1962; Shabansky, 1968).

A. Observations of the Interplanetary Medium

The term "interplanetary space" refers to the vast domain beyond the regions of the solar and the planetary atmospheres. No specific distinction will be made between the outermost layer of the solar atmosphere and the interplanetary gas. The unusual interest and attention given to the interplanetary medium result from the following reasons: many geophysical effects observed in the earth's atmosphere, such as aurora and geomagnetic storms, are influenced by the activities occurring and propagated in the interplanetary medium. Observations which provide evidence of the presence of the interplanetary medium and some of its properties will be briefly described.

The existence of an interplanetary medium was probably first indicated by the observation of the zodiacal light, or sunlight scattered by free electrons and dust particles filling the space between the planets (Behr and Siedentopf, 1953). Although at present the exact percentage of scattered sunlight attributed to the free electrons is still uncertain, it appears that the upper limit for the electron density at a distance of one astronomical unit from the sun is at least 400 electrons/cm^3; the particle density of dust might be in the range of 10^{-13}–10^{-15}/cm^3, depending on the material of the dust, according to Blackwell and Ingham (1961).

The next source of information on the interplanetary medium comes from observations of cometary tails. There are essentially two types of cometary tails according to their composition, and in any particular comet either or both types may be present. One of these consists mainly of dust particles; the other, plasma particles. A dusty comet is often curved and structureless; the accelerations of its dust particles can be considered as a consequence of the pressure of sunlight and the solar gravity. On the other hand a plasma comet is straight and points in a direction almost radially away from the sun when the comet approaches the sun to a distance within a few astronomical units. It was found that the accelerations in ionized tails are far too high to be attributed to sunlight pressure. This prompted Biermann (1951) to suggest the presence of a continual solar plasma stream flowing radially away from the sun, which interacts with the cometary plasma to provide the latter with the necessary momentum for high accelerations. Although Biermann's proposed microscopic mechanism of interaction between the solar and the cometary plasma particles appears insufficient to account for the observed cometary acceleration, his concept of continual solar stream is recognized as an important innovation to the development of the hydrodynamic theory of solar wind (Parker, 1958).

The use of comets, either natural or artificial, as probes of the interplanetary medium could prove highly rewarding in the fluid-mechanical approach to astrophysical problems. The gas-dynamical interaction between solar wind and a comet has many features in common with that of the wind with a planet which has negligible intrinsic magnetic field. The fact that the comets are accessible for observations is an important asset. In cometary studies, the search for the ionization mechanism primarily responsible for the ionized gases present in the cometary atmosphere has not yet been successful. This is relevant to the microscopic momentum transfer between the solar and the cometary particles and hence has important influence on the gas dynamics of a planet (Biermann *et al.*, 1957).

During the last decade it has been possible to make direct measurements of the interplanetary medium with the help of spacecraft and the earth's artificial satellites. One of the unique features in such *in situ* measurements in space is the direct determination of fluxes of the interplanetary ion species. This has never been possible to obtain from remote-sensoring observations where the consequences of electron activities are more pronounced. It is of interest to note that, among others, Parker's hydrodynamic theory of solar wind (Parker, 1958) has been shown to agree well with the experimental results from *in situ* measurements via satellites.

B. Structure of the Solar Atmosphere

It is clear from the above discussion that both the material and energy of the interplanetary gas originate primarily from the sun. To prepare for the later discussion of the dynamics of gases in the interplanetary space, a brief description of the solar energy source and the solar atmosphere is in order.

The sun is a gaseous body. The center of the sun is extremely hot, with a temperature of the order of 15×10^6 °K, and extremely dense, with a density 100 times that of water. According to Bethe (1939) the energy release from the fusion of hydrogen nuclei into helium near the center of the sun acts as the source of solar energy. The solar energy is transported outwards by the radiation processes. X-rays predominate at the center of the sun and it is by their slow diffusion and transfer outward in the direction of temperature descent that the energy finally reaches the 6000°K surface of the sun. The temperature profile in the cross section of the sun is shown in Fig. 1. Above the visible surface of the sun, the atmosphere extends out

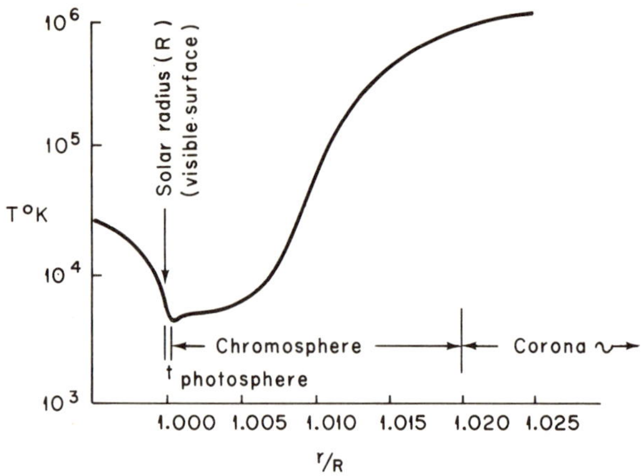

FIG. 1. Temperature distribution in solar atmosphere [based on data provided by Allen (1963, Chap. 9)].

for a considerable distance into the interplanetary space and beyond. The inner part of this atmosphere forms the solar corona, while the outer part is usually referred to as interplanetary gas. The solar atmosphere appears to consist mainly of hydrogen in a fully ionized state. It also contains a few percent of helium and a much smaller proportion of heavier elements. In what follows it will be assumed, for the sake of simplicity, that it consists entirely of hydrogen. Shown in Table 1 are crude average values, observed during the periods of quiet sun, of the physical quantities that characterize the properties of the solar atmosphere at various heliocentric distances (measured in terms of the solar radius $R = 7 \times 10^{10}$ cm).

Astronomical observations on the scattering of sunlight have provided estimates of the number density of electrons in the solar atmosphere up to about 10 solar radii. Recent space experiments give the number density of protons in the neighborhood of the earth's orbit, i.e. at a heliocentric distance of about 215 solar radii. The solar atmosphere is approximately electrically neutral, i.e. the protons and the free electrons have

TABLE 1[a]

r/R	1.03	1.5	3	5	10	215
$n(\text{cm}^{-3})$	2×10^8	1.6×10^7	3.8×10^5	3.6×10^4	4.1×10^3	7
$T(^\circ\text{K})$	1.7×10^6	10^6	7×10^5	5×10^5	4×10^5	4×10^4
B(gauss)	1	0.44	0.11	0.04	0.01	3×10^{-5}
V(km/sec)						3.6×10^2

[a] Rossi and Olbert, 1970, Chap. 14.

approximately the same number density (n). The temperature T has been estimated from a variety of astronomical data including profiles of emission lines, intensity of radio noises, etc. The values of B listed in Table 1 for $r < 10R$ have been computed from the magnetic field observed at the solar surface under the assumption of a $1/r^2$ radial dependence. The value of B at 215 R was derived from the satellite measurements; so was the solar wind velocity at 215 R.

It is of interest to note the rapid temperature rise in the chromosphere and corona of the sun away from the solar energy source. The nature of the heating appears incompatible with the second law of thermodynamics which says that in a closed thermodynamic system heat cannot of itself pass from a cooler to a hotter body. The obvious way out of this difficulty is to suggest that work must have been done as in a heat pump. This leads to the idea that motions taking place in the solar photosphere may be doing work on the atmosphere, transferring ordered mechanical energy in the form of waves propagating towards the upper atmosphere of the sun where the density is very low and there dissipating their energy in the form of heat. This mechanism of energy transport has often been referred to in cosmic gas dynamics.

C. Kinetic Properties of the Interplanetary Plasma

The discussion so far establishes that the interplanetary space has been found to be filled with a high temperature plasma that owes its origin primarily to the sun. In this huge volume of tenuous plasma surrounding the sun, fascinating and complex processes are expected to take place on account of the unusual kinetic behavior of charged particles imbedded in a magnetic field. The peculiarities of charged particles that have most important bearings on the macroscopic behavior of a plasma are the long-range nature of their interparticle forces and their persistent gyrations around magnetic field lines.

In the kinetics of the Boltzmann gas whose neighboring particles have a short-range interaction it is possible for a dilute gas to ignore the effect of multiple collision in comparison with that of the binary collisions. This is one of the basic hypotheses in the derivation of the Boltzmann equation which governs the change in the distribution of particles streaming in a six-dimensional phase space $(\mathbf{x}, \mathbf{v})$. On the other hand with long-range interaction of the charged particles which follows Coulomb's law, it can be shown that the hypothesis of dominance of binary collisions is invalid. If the space around a single charged particle is divided into spherical shells with radial distance (r) in equal increments, the number of particles in successive shells increases proportionally with r^2. At a given density, the number of particles increases with volume, i.e. with r^3. Since the forces exerted by these particles on the central one decreases with $1/r^2$, many shells may act simultaneously with equal force. In fact, with a typical plasma, the rare binary collisions have negligible influence compared with the cumulative multiple distant "collisions." This feature of charged particles gives rise to collective modes of plasma behavior which do not exist in gases of molecules with short-range interactions. This is why plasma interaction presents a many-body problem and causes great difficulties in the kinetic theory of plasma.

As a prelude to the discussion of contemporary approximations in the study of many-body effects in plasma kinetics, it is pertinent to explain the electric field shielding phenomenon in plasma which not only exemplifies the collective action of charged particles but also shortens the range of particle interaction from a pure Coulomb r-dependence into one which is modified by a cutoff factor due to the cumulative space charge effect. It was Debye who, in constructing a theory of strong electrolytes, discovered that around each charged particle in a nonmagnetic, macroscopically neutral plasma* there exists a neighborhood in which there is a greater probability of finding a particle of opposite charge than one of the same sign. This is due to Coulomb's law which says that charges of like sign repel each other and charges of unlike signs attract each other. The effect of this congregation of particles with opposite charges around a test particle tends to develop an electric field to neutralize that of the test particle. This leads to a potential field $\sim r^{-1} \exp(-r/\lambda_D)$ where $\lambda_D = (KT/4\pi ne^2)^{1/1}$ is known as the Debye shielding length. The shielding factor $\exp(-r/\lambda_D)$ makes the resultant field shorter in range as compared to the Coulomb field $(\sim 1/r)$. This microscopically nonhomogeneous particle distribution implies that in a small volume element the number of

* In a large volume containing many particles, the numbers of positive and negative charges are the same.

charges of either sign will fluctuate and neutrality will not hold except as an average over time or over space. In other words a macroscopically neutral plasma under statistical equilibrium exhibits microscopic nonhomogeneities of characteristic length λ_D. The spherical region in which the charge density is nonuniform and in which the electrostatic field differs appreciably from zero is called the Debye sphere, whose radius is equal to the Debye length λ_D. This shielding effect is responsible for the fact that an electrostatic field with a scale larger than λ_D cannot be maintained in a plasma. Static fields of external origin cannot penetrate deeper than a distance of the order of a Debye length. Rapidly oscillating fields, however, can exist.

II. Methods of Gas Kinetics and Continuum Flows

A fluid motion can be visualized as a collected motion of a large number of particles of which the fluid is kinetically composed. A mathematical representation of the fluid motion can be made at different levels of approximations: A microscopic approach to the flow problem starts with the application of the conservation laws of mechanics to the particle motions and the use of mathematical statistics to evaluate mean values of particle motions that are macroscopic observables of the flow field. On the other hand a macroscopic approach to the flow problem starts with the choice of a representative fluid element whose macroscopic behavior can be construed as giving the average behaviors of particles in the element. Laws of mechanics are then applied to the motion of the fluid element. The formal difference between these two approaches is merely a matter of procedure. There is, however, much subtlety in the choice between them for the solution of a flow problem.

A. Microscopic and Macroscopic Representations

The kinetic theory of gases was developed on the assumption that a gas consists of molecules in irregular or chaotic motions. As a first approximation, suitable for neutral particles, these molecules are considered as having a small but finite size which, for a dilute gas, is much less than the mean distance between the molecules. Thus, the molecules move freely through space, except for occasional collisions. These collisions produce abrupt changes, both in the direction and in the magnitude of the molecular velocity. It can be seen intuitively that in time a random distribution of particle positions and directions of motion is to be expected because of these collisions. Consider submicroscopic particles, e.g. smoke particles, suspended in the air and the effect of the irregular motions of air molecules on the smoke particles. Although each smoke particle is small, it still contains of the order of 10^8 atoms or more. Thus when it is struck by a molecule of the gas in which it is suspended, it will receive an impulse which causes it to change its velocity by only a small amount even though the gas molecules are moving quite rapidly ($\sim \frac{1}{2} \times 10^5$ cm/sec). Since it is being struck continually and in a very irregular way by the gas molecules, a corresponding slow but irregular fluctuation develops in the speed of the smoke particle in the manner of an almost continuous motion depending on the mass ratio of a molecule and the smoke particle. This is known as the Brownian motion in statistical physics.

The significance of the Brownian motion example is that the large-scale overall statistical regularities can appear at the macroscopic level, which are largely independent of the precise details of the complicated and irregular motions taking place at the molecular level. This kind of overall statistical law has become quite general, not only in physics but also in many other fields. Consider a box of gas containing molecules of the order of 10^{23}, each of which is moving in a very irregular path as a result of continual collisions with the other molecules. Clearly it is nearly impossible to solve for each molecular motion, not only because of the mathematics involved but also due to a lack of precise initial values of position and velocity of each molecule. Indeed there is the added complication of instability in the calculation of a molecular motion, e.g. a slight change of the initial angle of motion of any molecule will change appreciably its direction of motion after the first collision. This change will in turn lead to a still bigger change in the next collision, etc., and the cumulative effects of these changes will finally carry the molecule in question to a very different region of space and thus provide an effective randomization of the molecular motions.

It is noted that the very same factors which make a detailed trajectory calculation impossible are also those that make possible a general prediction of the macroscopic average properties of the system without the need for precise information about exactly what the individual molecules are doing. This insensitivity of the macroscopic average quantities, such as the mean velocity of the Brownian particles, the fluid shear stress on a unit surface, etc. to the precise motions and arrangements in space of the individual molecules originates, at least in part, in the fact that an enormous number of different motions and arrangements in space can lead to practically the same values for these quantities. Consequently if we interchange two molecules in space, we obtain a different set of microscopic conditions for the same macroscopic effects. Also if, in a given region of space, a given molecule changes its direction of motion, the effects of this change on large-scale averages can be well compensated by suitable opposite changes in the motions of other molecules in the sample for averaging. Thus irregular motions of many molecules will produce fluctuations, the effects of which tend, in the long run and on the average, to cancel out. Indeed, the above considerations have been verified quantitatively in studies of statistical mechanics which show that almost all possible initial conditions for the molecular positions and velocities lead to irregular motions in which the large-scale averages fluctuate very close to practically determinate mean values. Because these mean values depend almost entirely on the general overall properties of the molecules, such as the mean density, the mean kinetic energy, etc., which can be defined directly at the large-scale level, it becomes possible to obtain regular and predictable relationships involving the large-scale level alone, namely, a set of macroscopic causal laws of thermodynamics and hydrodynamics.

B. The Kinetic Flow Equations

Consider a mixture of ionized and possibly neutral particles. The statistical mechanical state of the gas mixture can be conveniently described by separate distribution functions each of which denotes a particular species of the mixture, e.g. $f_\alpha(\mathbf{x}, \mathbf{v}, t)\, d\mathbf{x}\, d\mathbf{v}$ represents the expected number of particles species α in the six-dimensional phase space element $d\mathbf{x}\, d\mathbf{v}$ at time t where α may be used to designate ions ($\alpha = i$), electrons ($\alpha = e$), or neutral ($\alpha = n$). The kinetic equation, also known as the Boltzmann equation, for the probability density $f_\alpha(\mathbf{x}, \mathbf{v}, t)$ governing the statistical distribution of species α which are streaming in the phase space $(\mathbf{x}, \mathbf{v})$ can be written (Chapman and Cowling, 1970, p. 361)

$$\frac{\partial f_\alpha}{\partial t} + \mathbf{v} \cdot \frac{\partial f_\alpha}{\partial \mathbf{x}} + \mathbf{a} \cdot \frac{\partial f_\alpha}{\partial \mathbf{v}} = \left(\frac{\delta f_\alpha}{\delta t}\right)_{\text{coll}}. \tag{2.1}$$

where $\mathbf{a}(\mathbf{x}, \mathbf{v}, t)$ is the particle acceleration due to the Lorentz force $q(\mathbf{E} + \mathbf{v} \times \mathbf{B}/c)$ on a particle of charge q and possibly the gravitational force if it is significant. The electrostatic field $\mathbf{E}$ is the self-consistent electrostatic field which takes account of the long-range collective part of the Coulomb interaction while the collision term $(\delta f_\alpha/\delta t)_{\text{coll}}$ accounts for the residual short-ranged (screened) interaction between charged particles. This *ad hoc* division of the many-body interaction effect into the long-ranged collective interactions and the short-ranged discrete interactions, and the subsequent assumption of the dominance of the former contribution, must be considered as one of the fundamental assumptions of contemporary plasma physics.

The self-consistent nature of the electromagnetic field is built into the equation when the field is required to satisfy Maxwell's field equations in which the current $\mathbf{J}$ is written

$$\mathbf{J} = \mathbf{J}_{\text{ext}} + \sum_{\alpha} q_\alpha \int f_\alpha(\mathbf{x}, \mathbf{v}, t)\mathbf{v}\, d\mathbf{v}, \tag{2.2}$$

where $\mathbf{J}_{\text{ext}}$ is any externally produced current and the second term corresponds to the current produced by the plasma itself. The summation is over various charged species that compose the plasma. A similar equation relates the charge density ρ_e to the distribution function:

$$\rho_e = \sum_{\alpha} q_\alpha \int f_\alpha\, d\mathbf{v}. \tag{2.3}$$

The approximation to the discrete-interaction effect, namely $(\delta f_\alpha/\delta t)_{\text{coll}}$, has been of considerable historical interest. In the binary collision analysis of a pair having the inverse-square law (Coulomb's law for charged particles) of interaction, difficulties arise because important contributions to the collision integral (Chapman and Cowling, 1970, p. 177) are made by encounters in which the mutual distance of the particles is always large; at the same time, a molecule at a large distance from a given molecule will also be under the Coulomb interactions of many other molecules, so that these "distant" encounters are not really binary. In fact, it is a feature of the Coulomb interaction that multiple scattering (many-body collisions) is much more important than single scattering (binary collision). By this we mean that the time required for a test particle to suffer a large deflection as a result of multiple scattering is much shorter than the average time before the particle undergoing a single collision produces a similar deflection. This circumstance justifies the application of the Fokker–Planck approach, a collision model used successfully for Brownian motion investigations, to the study of discrete collision effects in plasmas. It is thus seen that in contemporary approximations to plasma particle interactions, a working hypothesis has often been introduced which treats the interaction effect as contributed by two separate causes: (i) the self-consistent field, Eqs. (2.1), (2.2), and (2.3); (ii) the Fokker–Planck collisions due to neighboring particles within the Debye shielding length λ_D. Note that the Debye shielding effect of plasma reduces the effective range of the Coulomb interaction between two particles in empty space to that of the order of λ_D. Although the Debye shielding length in dilute plasma is large in comparison with atomic dimensions, which are relevant when large-angle Coulomb scattering is considered, it still may often be treated as infinitesimal in the description of large-scale phenomena. For instance, in interplanetary surroundings, the Debye shielding length is typically of the order of several meters or less which is small in comparison with the size of a fluid element chosen in the study of the macroscopic behavior of an interplanetary plasma. This relative smallness of the Debye shielding length implies automatically a relative shortness of the interaction time and is important in the Fokker–Planck approximation (Chandrasekhar, 1943). A test particle will complete its interaction with a target particle within a time interval $\Delta t \sim \lambda_D/v$. Typically with $\lambda_D \sim 1$ m and $v \sim 10$ km/sec, $\Delta t \sim 10^{-4}$ sec; this is a sufficiently short time interval to be treated as infinitesimal compared with the various relaxation time constants of the interplanetary system.

Consider the plasma particles undergoing random motions, similar to Brownian motion, as a result of a large number of small deflections. It is appropriate to describe such a stochastic process by a transition probability $\psi(\mathbf{v}, \Delta\mathbf{v})$ that a particle with velocity $\mathbf{v}$ acquires a velocity increment $\Delta\mathbf{v}$ in time Δt. For the Markoff (stochastic) process ψ does not depend explicitly on time and the distribution then evolves according to the relation

$$f_\alpha(\mathbf{x}, \mathbf{v}, t) = \int f_\alpha(\mathbf{x}, \mathbf{v} - \Delta\mathbf{v}, t - \Delta t)\psi(\mathbf{v} - \Delta\mathbf{v}, \Delta\mathbf{v})d(\Delta\mathbf{v}). \tag{2.4}$$

It is assumed that a time interval Δt exists which is long enough for a particle to suffer a large number of collisions, but short enough so that its velocity does not change much as shown above. Hence the discrete collision effect is given by

$$\left(\frac{\delta f_\alpha}{\delta t}\right)_{\text{coll}} = \lim_{\Delta t \to 0} \frac{[f_\alpha(\mathbf{x}, \mathbf{v}, t) - f_\alpha(\mathbf{x}, \mathbf{v}, t - \Delta t)]}{\Delta t}. \tag{2.5}$$

When the functions under the integral in (2.4) are expanded into the Taylor series, we have

$$f_\alpha(\mathbf{x}, \mathbf{v}, t) = \int d(\Delta\mathbf{v})\left\{f_\alpha(\mathbf{x}, \mathbf{v}, t)\psi(\mathbf{v}, \Delta\mathbf{v}) - \Delta t\psi \frac{\partial f_\alpha}{\partial t} - \Delta\mathbf{v} \cdot \left[\frac{\partial f_\alpha}{\partial \mathbf{v}} + f_\alpha \frac{\partial \psi}{\partial \mathbf{v}}\right]\right.$$
$$\left. + \tfrac{1}{2}\Delta\mathbf{v}\,\Delta\mathbf{v} : \left[\psi \frac{\partial^2 f_\alpha}{\partial \mathbf{v}\,\partial \mathbf{v}} + 2\frac{\partial f_\alpha}{\partial \mathbf{v}}\frac{\partial \psi}{\partial \mathbf{v}} + f_\alpha \frac{\partial^2 \psi}{\partial \mathbf{v}\,\partial \mathbf{v}}\right] + \cdots\right\}, \tag{2.6}$$

which, after the use of normalization condition for ψ, namely

$$\int \psi\, d(\Delta\mathbf{v}) = 1,$$

reduces to

$$\Delta t \left(\frac{\delta f_\alpha}{\delta t}\right)_{\text{coll}} = -\frac{\partial}{\partial \mathbf{v}} \cdot (\overline{\Delta\mathbf{v}} f_\alpha) + \frac{1}{2}\frac{\partial^2}{\partial \mathbf{v}\,\partial \mathbf{v}} : (\overline{\Delta\mathbf{v}\,\Delta\mathbf{v}} f_\alpha), \tag{2.7}$$

where

$$\overline{\Delta\mathbf{v}} = \int \psi(\mathbf{v}, \Delta\mathbf{v})\, \Delta\mathbf{v}\, d(\Delta\mathbf{v}),$$

$$\overline{\Delta\mathbf{v}\,\Delta\mathbf{v}} = \int \psi(\mathbf{v}, \Delta\mathbf{v})\, \Delta\mathbf{v}\, \Delta\mathbf{v}\, d(\Delta\mathbf{v}).$$

The Fokker–Planck collision term can be used after ψ is prescribed for an assumed model of the scattering process. Note that the collision term $(\delta f_\alpha/\delta t)_{\text{coll}}$ in (2.7) satisfies the conservations of number, momentum, and energy, so that (Burgers, 1969, p. 123)

$$\int (1, \mathbf{v}, v^2)\left(\frac{\delta f_\alpha}{\delta t}\right)_{\text{coll}} d\mathbf{v} = 0. \tag{2.8}$$

However, interplanetary gas is so rarefied that the collective effect of the distant particles is much greater than any systematic effect arising from the discrete collisions involving one, or a small number, of the nearby particles. Consequently the collisionless Boltzmann equation, with self-consistent field, often called the Vlasov equation, is used exclusively (Chapman and Cowling, 1970, p. 403):

$$\frac{\partial f_\alpha}{\partial t} + \mathbf{v} \cdot \frac{\partial f_\alpha}{\partial \mathbf{x}} + \mathbf{a} \cdot \frac{\partial f_\alpha}{\partial \mathbf{v}} = 0. \tag{2.9}$$

The self-consistent electromagnetic field, particularly the electrostatic part, which is induced by the presence of a net space charge (see Eq. 2.3) and current (see Eq. 2.2) is mostly responsible for the seemingly paradoxical quasi-continuum behavior of the collisionless interplanetary plasma. This will be further exposed in Section III. It should be emphasized that for gases consisting only of neutral particles (Sections VI and VII), the situation is quite different. In that case it is the influence of the distant particles which may be neglected in comparison with that of the nearby particles, to such an extent that no mutual interaction need be included in **a**. In practice, the only contribution to **a** is the external gravitational field, if any. Further, only binary collisions need to be included in $(\delta f/\delta t)_{\text{coll}}$ if necessary. These facts all follow from the short-range nature of the interparticle force of interaction, e.g. the Van der Waals force between two neutral molecules. Thus, the orbit of each particle may be regarded as a succession of free orbits punctured by very brief encounters, each involving only one other particle, with the velocity changing abruptly at each encounter. It is exclusively for the situation where the binary collision dominates that the Boltzmann equation was originally derived (Chapman and Cowling, 1970, p. 46).

C. Formal Solution of Vlasov Equation

The distribution $f(\mathbf{x}, \mathbf{v}, t)$ of a collisionless flow which has a given initial distribution $f_0(\mathbf{x}, \mathbf{v}, t_0)$ can be determined from (2.9) by the method of characteristics. Notice that a first-order partial differential equation like (2.9) has the subsidiary equations in Cartesian coordinates

$$\frac{dt}{1} = \frac{dx_1}{v_1} = \frac{dx_2}{v_2} = \frac{dx_3}{v_3} = \frac{dv_1}{a_1} = \frac{dv_2}{a_2} = \frac{dv_3}{a_3} \tag{2.10}$$

which define the characteristic curves in a seven-dimensional space $(\mathbf{x}, v, t)$. The general solutions to (2.9) can be formally prescribed as follows:

$$x_\beta = x_\beta(I_1, I_2, \ldots, I_6, t); \quad v_\beta = v_\beta(I_1, I_2, \ldots, I_6, t), \quad \beta = 1, 2, 3, \tag{2.11}$$

where $I_1, I_2, \ldots, I_6$ are the six constants of integration. Notice that (2.10) are the equations of particle orbits and the solutions (2.11) represent the position and velocity of a particle at time t. The six constants $I_1, I_2, \ldots, I_6$ serve as constraints of the particle trajectory. It is formally feasible to rewrite relations (2.11) in the form

$$I_\beta = I_\beta(\mathbf{x}, \mathbf{v}, t), \tag{2.12}$$

where subscripts $\beta = 1, 2, \ldots, 6$. The parameters I_β are used here to specify a particular orbit passing through $\mathbf{x}$ with velocity $\mathbf{v}$ at time t. The general solution of (2.9) can be written as an arbitrary function $F(I_1, \ldots, I_6)$, with

$$f(\mathbf{x}, \mathbf{v}, t) = F(I_1, I_2, \ldots, I_6). \tag{2.13}$$

The particular solution to the given initial problem is obtained by demanding that $F(I_1, I_2, \ldots, I_6)$ be compatible with the given initial distribution $f_0(\mathbf{x}, \mathbf{v}, 0)$, using the trajectory relations (2.11) and (2.12).

A result similar to (2.13) had been obtained by Jeans in his study on stellar dynamics (Chandrasekhar, 1942, p. 131) and is known as Jeans' theorem, although in a somewhat different connection the theorem was also proved earlier by Poincaré. Extensive applications of (2.13) having selected constants of A_β in (2.12) identified with adiabatic invariants of ionic motions in an electrostatic field have been made in ioniospheric gas dynamics involving satellites and diagnostic probes (Liu, 1969).

It is of interest to note that in studying the nonlinear particle–field interaction in collisionless plasma flows with a self-consistent field, there is some mathematical advantage in the use of the wave mechanical formulation, after Schroedinger, for the particle motion (Liu, 1965, 1969; Harris, 1969).

D. Continuum Flow Equations

In view of the discussion in Section II,A, and of the fact that the macroscopic observables of a flow field are moments of the particle velocity distribution function $f(\mathbf{x}, \mathbf{v}, t)$, it is not surprising that sometimes a macroscopic approach to a flow problem is fruitful. The number density $n(\mathbf{x}, t,)$ the hydrodynamic velocity $\mathbf{u}(\mathbf{x}, t)$, and the pressure tensor $p_{\beta\gamma}(\mathbf{x}, t)$ are

$$n(\mathbf{x}, t) = \int f(\mathbf{x}, \mathbf{v}, t)\, d\mathbf{v}, \tag{2.14}$$

$$\mathbf{u}(\mathbf{x}, t) = \frac{1}{n} \int \mathbf{v} f(\mathbf{x}, \mathbf{v}, t)\, d\mathbf{v}, \tag{2.15}$$

and

$$\begin{aligned} p_{\beta\gamma}(\mathbf{x}, t) &= \int m(v_\beta - u_\beta)(v_\gamma - u_\gamma) f(\mathbf{x}, \mathbf{v}, t)\, d\mathbf{v} \\ &= m \int v_\beta v_\gamma f\, d\mathbf{v} - mnu_\beta u_\gamma , \end{aligned} \tag{2.16}$$

respectively, with m denoting the particle mass.

It is assumed that in the above averaging over the full range of v, the flow medium is treated as a quasi-continuum such that a meaningful fluid element in the physical space ($\mathbf{x}$) can be defined. It is found that by taking appropriate moments of (2.9) a system of conservation equations of flow can be obtained which can be solved directly for the macroscopic observables defined in relations (2.14)–(2.16), provided the conversation equations thus obtained can be made a closed set.

The zeroth moment of (2.9) can be written as

$$\int \frac{\partial f}{\partial t}\, d\mathbf{v} + \int \frac{\partial}{\partial x_\beta} (v_\beta f)\, d\mathbf{v} + \int \frac{\partial}{\partial v_\beta} (a_\beta f)\, d\mathbf{v} = 0. \tag{2.17}$$

The first term can be simplified by interchanging the order of the differential and the integral operator, leading to $\partial n/\partial t_\beta$ the second term becomes $\partial(nu_\beta)/\partial x_\beta$. The last term vanishes when the divergence theorem in the velocity space is used and if it is assumed that af approaches zero sufficiently fast as $\mathbf{v} \to \infty$. Thus we obtain the hydrodynamic continuity equation

$$\frac{\partial n}{\partial t} + \frac{\partial(nu_\beta)}{\partial x_\beta} = 0. \tag{2.18}$$

The first moment of (2.9) is

$$\int v_\beta \frac{\partial f}{\partial t}\, d\mathbf{v} + \int v_\beta \frac{\partial(v_\gamma f)}{\partial x_\gamma}\, d\mathbf{v} + \int v_\beta \frac{\partial(a_\gamma f)}{\partial v_\gamma}\, d\mathbf{v} = 0, \tag{2.19}$$

which, after simplification, becomes

$$\frac{\partial(nu_\beta)}{\partial t}+\frac{\partial}{\partial x_\gamma}\int v_\beta v_\gamma f\,d\mathbf{v}-\int a_\beta f\,d\mathbf{v}=0. \tag{2.20}$$

The use of relations (2.16), (2.18), and a definition of the total external force per unit volume $\mathbf{F}$ in (2.20) leads to

$$\frac{\partial u_\beta}{\partial t}+u_\gamma\frac{\partial u_\beta}{\partial x_\gamma}=\frac{1}{nm}\left(-\frac{\partial p_{\beta\gamma}}{\partial x_\gamma}+F_\beta\right), \tag{2.21}$$

which is the hydrodynamic momentum equation.

It is expected that a second moment of (2.9) will generate a higher order tensor, e.g. the heat flux tensor. The system of moment equations for the macroscopic observables like $n(\mathbf{x}, t)$, $\mathbf{u}(\mathbf{x}, t)$, $p_{\beta\gamma}(\mathbf{x}, t)$, etc., is not a closed set ready for solution. An additional hypothesis pertaining to the fluid is needed to complete the set. Under the condition that the molecular distribution is locally Maxwellian which implies isentropy, the pressure tensor becomes isotropic $p_{\beta\gamma}=p\,\delta_{\beta\gamma}$ where the thermodynamic pressure $p=p_{\beta\beta}/3=m\int(\mathbf{v}-\mathbf{u})^2 f\,d\mathbf{v}=nKT$. Substitution of $p_{\beta\gamma}=p\,\delta_{\beta\gamma}$ into (2.21) in the absence of $\mathbf{F}$ leads to

$$\frac{dp}{\partial t}+\frac{5}{3}\rho\frac{\partial u_\beta}{\partial x_\beta}=0, \tag{2.22}$$

which can be combined with the continuity equation (2.18) to give

$$\frac{d}{dt}\left(\frac{p}{n^{5/3}}\right)=0, \tag{2.23}$$

an isentropic relation for a gas with the ratio of specific heats of 5/3.

Fluid dynamicists will readily notice that the microscopic equation (2.9) bears the same relationship to the moment equations (2.18), (2.21), etc., as Prandtl's boundary-layer equation bears to von Kármán's momentum integral equation. It is recalled that the flow variables in Prandtl's equation describe the boundary-layer field pointwise while von Kármán's integral, being a moment of the boundary-layer equation, satisfies the momentum conservation on the average across the boundary layer only. The relaxed condition imposed in the latter approach makes its solution untenable for a unique pointwise description of the flow field. The loss of accuracy in flow-field description is compensated by its mathematical simplicity and versatility. The comparison of the cited alternative approaches to the boundary-layer problem is very much in parallel with that of the two alternative approaches to gas-kinetic problems. Of course the moments are taken in the velocity space instead. While the microscopic equation (2.9) gives the pointwise description in the μ-space $(\mathbf{x}, \mathbf{v})$, the continuum equations obtained via the averagings of the moments of (2.9) satisfy the gas kinetics on the average in the velocity space. It is noted however that a closed set of continuum equations is not expected as a result of the averaging processes. For instance, an $(n+1)$th moment is usually generated in an nth-order moment equation. The customary practice to close the set is to introduce a particular constitutive relation between the stress tensor and the rate of strain from physical considerations. Under this development, the momentum conservation equation for a Newtonian fluid satisfying Stokes viscosity relation leads to the continuum hydrodynamic equation of Navier-Stokes (Liepmann and Roshko, 1957, p. 336). The assumption of isotropic pressure tensor used herein earlier is essentially a statement of nonviscous flow.

Thus an inherent relation between the gas-kinetic and the continuum formulations of a flow problem exists. It is useful to have a criterion for the validity of the continuum approach which is usually preferred because of its mathematical simplicity and directness. The key step in relating the continuum equations to the kinetic equation is the averaging over velocity space which is meaningful only if a fluid element which gives representative local average molecular dynamics is available. For example, it would not be meaningful to define a "representative fluid element" of a flow field near a body of size (L) which is much smaller than the (binary) collisional mean free path of the molecules reflected from the body. Later in the discussion (see Section III) it will be shown that with a cosmic plasma where the collective behaviors of the charged particles are important the particle (binary) collisional mean free path may not be an appropriate measure of the continuum of a fluid under certain conditions.

III. Collective Particle Behavior of the Interplanetary Gas

Early works on the dynamics of rarefied plasma disclosed some apparently paradoxical results; for example in spite of the fact that the mean free path (based on binary collisions) of an ion in the solar wind at the earth orbit is large compared with the scale of the solar system, the coronal expansion behaves like that of a continuum, as demonstrated in the close agreement of results between *in situ* space observations and the hydrodynamic theory of solar wind in interplanetary space. Furthermore the interaction of the solar wind with the magnetosphere of the earth appears to effect a bow shock in front of the magnetosphere much like that of a blunt body placed in a supersonic flow. A "collisionless" shock transition seemingly required by the large mean free path is difficult to reconcile with the Boltzmann kinetics where binary collisions are considered responsible for conversion of the energy of orderly motion into that of random motions. In the laboratory the Langmuir paradox which refers to the near equilibrium distribution of electrons in an apparently collisionless discharge column has been well known to the plasma physicists. A satisfactory explanation of the above-cited paradoxical phenomena which show continuum-like behavior for a collisionless plasma is important not only to the study of interplanetary gas dynamics but also to the foundation of plasma physics. One of the consequences is the justifiable use of hydrodynamic formulations in the cosmic plasma dynamics (see Section IV).

It has been frequently shown that many relaxation processes in a rarefied plasma involve cooperative plasma oscillations that result from various modes of plasma microinstabilities. The wave-particle resonant interaction thereby effected could cause plasma randomization and hence transport fluxes without participation of binary collisions as in the kinetic process of Maxwell–Boltzmann. The plasma wave motion sustained by the self-consistent field effect well exemplifies the collective particle behavior in the coherent motion of many particles due to their coupling with the induced field. This implies that each particle is constrained by a common potential field which represents the average field generated by all the particles in the medium. The collective behavior, on which the quasi-continuum assumption for the interplanetary plasma rests, brings about the equivalence of the microscopic (particle) and the macroscopic (continuum) approaches to the large-scale interplanetary gas dynamics provided microinstabilities and their consequential wave-particle resonant interactions prevail. In order to substantiate the qualitative discussion here, we shall present a quantitative theory on the wave mechanism in a collisionless plasma. This will consist of two parts: (a) the wave dispersion which describes its dissipation (damping) or amplification (antidamping), (b) wave-particle resonant interaction. The former is treated in the framework of linear theory which is adequate when the amplitudes of the disturbed oscillations of particles and field are small. The latter, however, must be dealt with, a quasi-linear analysis, at least. It is noted that to examine completely a collisionless unstable plasma, such as is being done here, from the time of the onset of the instability to the final quiescent equilibrium state, four consecutive phases of development must be considered. During the first phase when the exponential growth of certain unstable wave modes develops, the linearized theory applies. In the second phase on account of wave-particle resonant interaction, the background particle distribution diffuses in such a way as to bring the growth rate of the unstable waves to zero, leaving a quasi-stationary spectrum of waves. In the third phase the waves in the quasi-stationary spectrum interact in such a way as to distort the spectrum but to keep the energy in the spectrum roughly constant. In the fourth phase the waves interact to produce virtual waves which in turn lose energy to the bulk of the particles by Landau damping. This nonlinear Landau damping produces a slow decay of the wave energy and a slow heating of the plasma, together with a diffusion of the plasma until all of the wave energy has been given up to the particles and the plasma is in a quiescent, stable state (Kadomtsev, 1965). To treat the interaction process occurring in the second phase a quasi-linear theory can be used with caution. The phases III and IV which occur roughly simultaneously can only be handled with complicated nonlinear analysis. The present discussion is restricted to the first and second phases. It is believed that the omission of the discussion pertaining to the third and fourth phases, about which our understanding is still incomplete, would not hinder the illustration as to how the quasi-continuum characteristics of a rarefied plasma is developed. Furthermore the discussion considers only the wave mode of longitudinal plasma oscillations and their damping in the absence of external magnetic field.

A. Longitudinal Waves in a Collisionless Plasma

Consider a rarefied plasma consisting of electrons and ions. Under the collisionless state, the distribution functions of electrons $f_e(\mathbf{x}, \mathbf{v}, t)$ and of ions $f_i(\mathbf{x}, \mathbf{v}, t)$, can be written as follows according to (2.9):

$$\frac{\partial f_e}{\partial t} + \mathbf{v} \cdot \frac{\partial f_e}{\partial \mathbf{x}} - \frac{e}{m_e} \mathbf{E} \cdot \frac{\partial f_e}{\partial \mathbf{v}} = 0, \tag{3.1}$$

$$\frac{\partial f_i}{\partial t} + \mathbf{v} \cdot \frac{\partial f_i}{\partial \mathbf{x}} + \frac{e}{m_i} \mathbf{E} \cdot \frac{\partial f_i}{\partial \mathbf{v}} = 0, \tag{3.2}$$

where m_e denotes electron mass, m_i the ion mass, $q = -e$ (electron), $q = e$(ion). The self-consistent field $\mathbf{E}$ is governed by the Poisson equation

$$\nabla \cdot \mathbf{E} = 4\pi n e \int (f_i - f_e)\, d\mathbf{v} \tag{3.3}$$

and $\nabla \times \mathbf{E} = 0$, since longitudinal waves alone are considered.

Limiting the discussion to oscillations of small amplitude, it is possible to linearize the above system of equations by letting

$$f_e(\mathbf{x}, \mathbf{v}, t) = f_e^{(0)}(\mathbf{v}) + f_e^{(1)}(\mathbf{x}, \mathbf{v}, t)$$

and

$$f_i(\mathbf{x}, \mathbf{v}, t) = f_i^{(0)}(\mathbf{v}) + f_i^{(1)}(\mathbf{x}, \mathbf{v}, t),$$

where

$$f_{\mathrm{e}}^{(0)} = (\sqrt{\pi} A_{\mathrm{e}})^{-3} \exp - (\mathbf{v} - \mathbf{u}_{\mathrm{e}})^2 / A_{\mathrm{e}}{}^2$$

and

$$f_{\mathrm{i}}^{(0)} = (\sqrt{\pi} A_{\mathrm{i}})^{-3} \exp - (\mathbf{v} - \mathbf{u}_{\mathrm{i}})^2 / A_{\mathrm{i}}{}^2$$

denote the equilibrium distribution of electrons and ions respectively,

$$A_{\mathrm{e}} = (2KT_{\mathrm{e}}/m_{\mathrm{e}})^{1/2}, \qquad A_{\mathrm{i}} = (2KT_{\mathrm{i}}/m_{\mathrm{i}})^{1/2},$$

$f_{\mathrm{e}}^{(1)}$ and $f_{\mathrm{i}}^{(1)}$ denote deviations from the respective equilibrium distribution assuming $f_{\mathrm{e}}^{(1)} \ll f_{\mathrm{e}}^{(0)}$ and $f_{\mathrm{i}}^{(1)} \ll f_{\mathrm{i}}^{(0)}$. The substitution of the above representations for f_{e} and f_{i} in (3.1)–(3.3) and neglect of terms of the second and higher orders in $f^{(1)}$ will lead to a system of linear equations which is suitable for the Fourier-Laplace transform:

$$f_{\mathrm{e}}^{(1)}(\mathbf{k}, \mathbf{v}, \omega) = \int_0^{\infty} dt \int \exp[-i(\mathbf{k} \cdot \mathbf{x} - \omega t)] f_{\mathrm{e}}^{(1)}(\mathbf{x}, \mathbf{v}, t)\, d\mathbf{x}, \tag{3.4}$$

$$f_{\mathrm{i}}^{(1)}(\mathbf{k}, \mathbf{v}, \omega) = \int_0^{\infty} dt \int \exp[-i(\mathbf{k} \cdot \mathbf{x} - \omega t)] f_{\mathrm{i}}^{(1)}(\mathbf{x}, \mathbf{v}, t)\, d\mathbf{x}. \tag{3.5}$$

Let g_{e} (or g_{i}) denote the perturbation $f_{\mathrm{e}}^{(1)}$ (for $f_{\mathrm{i}}^{(1)}$) at time $t = 0$; then the transform of the linearized equations (3.1)–(3.3) gives

$$f_{\mathrm{e}}^{(1)}(\mathbf{k}, \mathbf{v}, \omega) = i\left[g_{\mathrm{e}}(\mathbf{k}, \mathbf{v}) + \left(\frac{eE}{m_{\mathrm{e}}}\right) \cdot \frac{\partial f_{\mathrm{e}}^{(0)}}{d\mathbf{v}}\right] \Big/ (\omega - \mathbf{k} \cdot \mathbf{v}), \tag{3.6}$$

$$f_{\mathrm{i}}^{(1)}(\mathbf{k}, \mathbf{v}, \omega) = i\left[g_{\mathrm{i}}(\mathbf{k}, \mathbf{v}) - \left(\frac{eE}{m_{\mathrm{i}}}\right) \cdot \frac{\partial f_{\mathrm{i}}^{(0)}}{\partial \mathbf{v}}\right] \Big/ (\omega - \mathbf{k} \cdot \mathbf{v}), \tag{3.7}$$

and

$$\mathbf{E}(\mathbf{k}, \omega) = -4\pi n_0\, e\, i(\mathbf{k}/k^2) \int (f_{\mathrm{i}} - f_{\mathrm{e}})\, d\mathbf{v}, \tag{3.8}$$

which, after the substitution of (3.6) and (3.7), leads to

$$\varepsilon(\mathbf{k}, \omega) \mathbf{E}(\mathbf{k}, \omega) = \frac{4\pi n_0 e}{k^2} \mathbf{k} \int \frac{g_{\mathrm{i}} - g_{\mathrm{e}}}{\omega - \mathbf{k} \cdot \mathbf{v}}\, d\mathbf{v}, \tag{3.9}$$

where

$$\varepsilon(\mathbf{k}, \omega) = 1 - \frac{\omega_{\mathrm{e}}{}^2}{k^2} \int \mathbf{k} \cdot \frac{\partial f_{\mathrm{e}}^{(0)}/\partial \mathbf{v} + (m_{\mathrm{e}}/m_{\mathrm{i}})\, \partial f_{\mathrm{i}}^{(0)}/\partial \mathbf{v}}{\mathbf{k} \cdot \mathbf{v} - \omega} \tag{3.10}$$

denotes the dielectric function for a two-component plasma.

The inverse Laplace transform

$$\mathbf{E}(\mathbf{k}, t) = (2\pi)^{-1} \int_{-\infty}^{\infty} \mathbf{E}(\mathbf{k}, \omega) \exp(-i\omega t)\, d\omega, \tag{3.11}$$

where the contour of integration must pass above all singularities of $\mathbf{E}(\mathbf{k}, \omega)$ since for $t > 0$, it can be closed below. The contributions from each singularity are to be added. Those arising from the right-hand side of (3.9) depend obviously on the particular initial conditions. On the other hand $\varepsilon(\mathbf{k}, \omega)$ depends only upon the equilibrium distributions $f_e^{(0)}$ and $f_i^{(0)}$. The relation

$$\varepsilon(\mathbf{k}, \omega) = 0 \tag{3.12}$$

is usually called the dispersion equation of the plasma. With the mean electron velocity $\mathbf{u}_e$ and the mean ion velocity $\mathbf{u}_i$ both parallel to the x axis, relation (3.12) can be written (Fried and Gould, 1961)

$$\varepsilon = 1 - \frac{k_D{}^2}{2k^2}\left[Z'\left(\frac{\omega - ku_e}{kA_e}\right) + \frac{T_e}{T_i} Z'\left(\frac{\omega - ku_i}{kA_i}\right)\right], \tag{3.13}$$

where $k_D = (4\pi ne^2/KT)^{1/2}$, the function Z is defined as

$$Z(\zeta) = 2i \exp(-\zeta^2) \int_{-\infty}^{i\zeta} \exp(-t^2)\, dt$$

$$= \exp(-\zeta^2)[i\sqrt{\pi} - 2 \int_0^{\zeta} \exp(s^2)\, ds, \tag{3.14}$$

and $Z'(\zeta) = dZ/d\zeta$.

A considerable amount of discussion is available in the publications on plasma waves pertaining to the search for the roots of (3.13) (see, e.g., Stix, 1962). The relevant results can be summarized under three distinguished cases as follows:

1. *Stationary Ions* ($m_e/m_i \to 0$) *and Zero Drift* ($u_e = u_i = 0$)

Under these conditions, all roots of (3.13) lie in the lower half ω plane implying damped oscillations; the least damped root ($\omega = \omega_r - i\gamma$), for $k \ll k_D$, gives with the plasma frequency denoted by $\omega_e = (4\pi n_0 e^2/m_e)^{1/2}$,

$$\omega_r = \left(\omega_e{}^2 + \frac{3k^2A_e{}^2}{2}\right), \qquad \gamma = \sqrt{\pi}\,\omega_e\left(\frac{\omega_e}{kA_e}\right)^3 \exp\left(-\frac{3}{2} - \frac{k_D{}^2}{2k^2}\right) \tag{3.14}$$

for the real and imaginary parts of the complex frequency ω. The corresponding dielectric function $\varepsilon(k, \omega)$ thus contains an imaginary part which implies the existence of damping action whose magnitude is determined by, among other factors, the derivative $\partial f^{(0)}/\partial v$ evaluated at $v = \omega_r/k$, where the particle velocity (v) equals the phase velocity of the wave. This phenomenon is known as Landau damping since it was discovered by Landau (1946). The Landau damping has been verified experimentally

(Malmberg and Wharton, 1964). Landau damping develops from wave-particle resonance. It is a damping if wave energy is absorbed by the resonant particles and an amplification (or negative damping) if particle thermal energy feeds the growing waves. The latter is often referred to as a trapping instability because the resonant electrons are trapped in the electric field of the waves for energy transfer to be effected. The Landau-like dampings of other plasma wave modes, e.g. transverse waves, have also since been developed (Montgomery and Tidman, 1964, p. 148).

2. *Plasma Ion Oscillations (m_e/m_i Small but Nonzero with Zero Drift, $u_i = u_e = 0$)*

When the first-order effect of m_e/m_i is included, a new set of roots of (3.13) appears with frequencies smaller than the plasma electron oscillation by a factor of $(m_e/m_i)^{1/2}$ and gives the plasma ion oscillation (ω_i). It has been shown that if the electron and ion temperature are equal ($T_e = T_i$), then even the least damped of these ion oscillations has $|\gamma| \geq |\omega_r|$ for all k, including $k = 0$, in contrast to the results (3.14). On the other hand, if $T_i < T_e$, which is common in many interplanetary conditions (Sturrock and Hartle, 1966), then the damping decreases and in the limiting case of $T_i \ll T_e$ with $k/k_D \ll (T_e/T_i)^{1/2}$, it was shown (Fried and Gould, 1961)

$$\omega = \left(\frac{m_e}{2m_i}\right)^{1/2} kA_e\left(1 + \frac{k^2}{k_D{}^2}\right)^{-1/2}\left[1 - i\left(\frac{\pi m_e}{m_i}\right)^{1/2} \frac{1}{2^{3/2}}\left(1 + \frac{k^2}{k_D{}^2}\right)^{-3/2}\right]. \tag{3.15}$$

It is of interest that for $k \ll k_D$ (ion Debye wave number), these ion waves have a constant phase velocity $\omega/k = (KT_e/m_i)^{1/2}$ which is of the same order of magnitude as acoustic wave velocity in a medium with $T_e = T_i = T$. Hence the term ion-acoustic wave is often used for the longitudinal ion waves. It is noted however that the restoring forces for the ion oscillations are electrical, arising from electron–ion charge separation, rather than mechanical, arising from the collisional momentum transfer that gives rise to ordinary acoustic waves in a collisional neutral gas. The ion-acoustic waves have been shown to play an important role in some collisionless shock transitions, e.g. the bow shock in front of the magnetosphere (Tidman and Krall, 1971).

3. *Bunching Phenomenon*

When a relative drift of electrons and ions exists [current $J = ne(u_i - u_e)$], it is found that the damping of the plasma ion waves decreases with increasing J until a critical value is reached where the imaginary part of ω

becomes positive, implying growing waves. If $T_e = T_i$, the minimum velocity for which there are growing waves is, according to Fried and Gould (1961),

$$u_i - u_e = 0.925[1 + (m_e/m_i)^{1/2}]A_e. \tag{3.16}$$

The least stable waves are those with $k = 0$. With $T_e > T_i$, the critical current becomes smaller. In the limit of $T_e \gg T_i$ the criterion for wave growth is

$$u_i - u_e \geq 4A_i. \tag{3.17}$$

The above instability is often known as the two-stream instability which occurs frequently in the interplanetary gas dynamics.

The physical mechanism responsible for the wave growth is the density modulations which arise from a space-varying electric field causing velocity modulations of the particles and conservation of particles ($nv =$ constant). If the electric field due to density variations is in phase with the initial field, then wave amplification results. In the extreme case, $T_e = T_i = 0$, any finite drift velocity (u_d) will cause wave growth with

$$k \leq (\omega_e/u_d)[1 - (m_e/m_i)^{3/2}]^{3/2}. \tag{3.18}$$

With finite temperatures, Landau damping causes wave decay.

B. Wave-Particle Resonant Interaction

The linear theory of plasma microinstabilities, as developed in the above discussion, is valid for disturbances of infinitesimally small amplitude only. Although it helps in establishing a necessary condition for trapping instability, namely $(\partial f_e^{(0)}/\partial v)_{v=\omega_r/k} > 0$, it is inadequate for treating the development of wave growth due to energy transfer by the wave-particle resonant interaction which is a nonlinear effect. To circumvent the difficulty of a nonlinear analysis, a quasi-linear theory of resonant interaction (Drummond and Pines, 1962; Vedenov, *et al.*, 1962) was proposed for the case where $\gamma \ll \omega_r$, under which there exists two distinct time scales, γ^{-1} and ω_r^{-1}. The basic approximation is that the space independent of the distribution function which in linear theory is $f_e^{(0)}(\mathbf{v})$, is now a function of time only on the slow time scale, whereas the electric field of the unstable plasma and the space dependent part of f_e depend on both time scales. In the one-dimensional electron oscillations (along the x axis) among cool ions assuming a small velocity bump which signifies a superposed electron beam (Fig. 2) the electron distribution function may be written

$$f_e(x, v, t) = f_e^{(0)}(v, t) + f_e^{(1)}(x, v, t), \tag{3.19}$$

where the average

$$\langle f_e^{(1)} \rangle = \int_{\infty}^{\infty} f_e^{(1)}(x, v, t)\, dx = 0.$$

The substitution of (3.19) in the Vlasov equation (2.9) leads to

$$\partial f_e^{(0)} / \partial t = -(e/m_e) \langle E\, \partial f_e^{(1)} / \partial v \rangle \tag{3.20}$$

and

$$\frac{\partial f_e^{(1)}}{\partial t} + v \frac{\partial f_e^{(1)}}{\partial x} - \frac{e}{m_e} E \frac{\partial f_e^{(0)}}{\partial v} = -\frac{e}{m_e} \left[E^{(1)} \frac{\partial f_e^{(1)}}{\partial v} - \left\langle E^{(1)} \frac{\partial f_e^{(1)}}{\partial v} \right\rangle \right]. \tag{3.21}$$

The terms on the right-hand side of (3.21) represent the mode coupling effect which is small at least for the simplest case of a small velocity bump (Fig. 2) in the trail of a Maxwellian distribution.

In a quasi-linear analysis, these coupling terms will be neglected in the initial iteration of (3.21); consequently the iterated equation (3.21) can be solved for $f_e^{(1)}$ whose primary time dependence is on the fast time scale (ω_r^{-1}), treating $f_e^{(1)}$ as essentially independent of time. The resulting $f_e^{(1)}$ thus obtained is substituted in (3.20) to give (Kadomtsev, 1965, Chap. 1)

$$\frac{\partial f_e^{(0)}}{\partial t} = \frac{\partial}{\partial v} \left(D_v \frac{\partial f_e^{(0)}}{\partial v} \right), \tag{3.22}$$

where

$$D_v = \frac{e^2}{m_e^2} \int \frac{\gamma E^2{}_k(t)}{(\omega - kv)^2 + \gamma^2}\, dk,$$

and E_k denotes the amplitude of the electric field E_k in the kth mode of Fourier resolution. For a constant mean distribution function the dependence of E_k^2 on t would be simply exponential, that is, $E_k^2 \sim e^{2\gamma t}$; for a weak dependence of $f_e^{(0)}$ on t, E_k^2 is given by

$$dE_k^2/dt = 2\gamma E_k^2. \tag{3.23}$$

Equations (3.22) and (3.23) constitute the basic equations of quasi-linear theory of wave-particle resonant interaction.

It is of interest to review the essential physics of quasi-linear theory which treats the electric field in an unstable plasma that has grown to such an amplitude whereby its reaction on the individual particle motion is no longer negligible. The diffusion coefficient D_v, initially insignificantly small because E_k is initially small will also grow in time, eventually causing nonnegligible diffusion in velocity space which, in turn, tends to reduce $\partial f_e^{(0)}/\partial v$ at $v = \omega_r/k$ and hence reduces γ and thus reduces the time growth of E_k^2. Thus the initial bump in Fig. 2 will gradually become

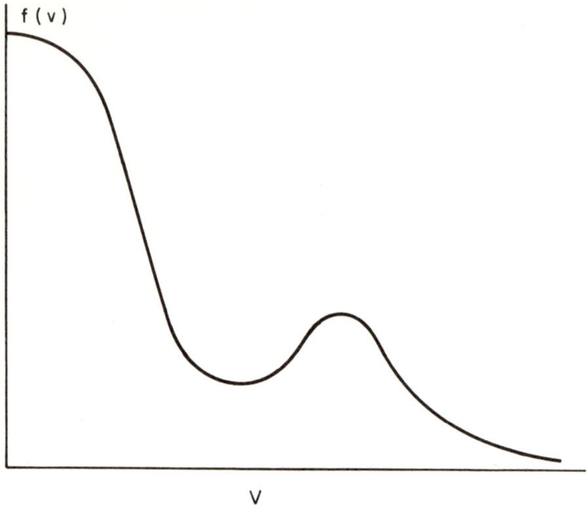

FIG. 2. Initial velocity distribution.

flattened. It is anticipated that a final steady state is possible in which E_k^2 is nonzero for a range of initially unstable modes, a state of a weakly turbulent plasma, namely with stationary electric fluctuations associated with the equilibrium wave spectrum which is intrinsically different from the ordinary hydrodynamic turbulence (Batchelor, 1953). It should be kept in mind that a quasi-linear theory as such is only applicable for times of an order less than ω_r^{-1}. For longer times the theory must be extended, e.g. with the inclusion of mode-coupling terms. Nevertheless the basic idea of a diffusion-like process in velocity space which tends to limit the initial wave growth appears plausible.

The extension of the present idea to include the presence of an external uniform magnetic field has also been made (Kadomtsev, 1965, Chap. 1). A quasi-diffusion equation in two dimensions ($v_\perp$ and $v_\parallel$) is obtained. In conclusion, the present analysis, although incomplete as far as predicting the final state of an unstable plasma is concerned, shows the presence of nonzero fluctuating electric and magnetic fields. Pertaining to individual particle motion the presence of these fluctuating fields acts somewhat analogous to collisions. Therefore, from the macroscopic point of view one should expect enhanced resistivity and diffusion of particles, in other words, a quasi-continuum character for the medium.

It is of interest to note that with the presence of an external magnetic field, more modes of microinstabilities are expected. Particular attention is drawn to the universal instabilities which is a low-frequency drift instability that has its origin in the diamagnetic current inherent to an inhomogeneous magnetic field (Krall, 1968). The effect of temperature gradient on the drift instabilities in an inhomogeneous high-β plasma (which is found in interplanetary space) is shown to cause ion-acoustic and Alfvenic instabilities (Hung and Liu, 1971).

IV. Hydrodynamic Coronal Expansion

When the solar atmosphere streams away from the sun into the interplanetary space the (binary) collisional mean free path (l) must rapidly increase so that near the earth orbit, it is found that l would be of the order of 1 AU($=1.5 \times 10^{11}$ m). This naturally caused some concern about the use of continuum fluid equations beyond the dense collisional region of the corona in the minds of earlier workers. In view, however, of the fact (see Section III) that some microscopic plasma instabilities and their associated waveparticle scattering must produce a thermalization mechanism leading to effective fluid transport characteristics in the interplanetary gas, it appears justifiable to formulate the large-scale coronal expansion flow hydrodynamically. To a large extent the present understanding of the local properties of the solar wind is based on analytical studies carried out in the manner of continuum fluid dynamics (Parker, 1958). These calculations have provided reasonably accurate theoretical values of the bulk or average flow properties near the earth orbit as compared with *in situ* measurements via spacecraft (Hundhausen, 1968). It should be warned that, in spite of the success of hydrodynamic theory pertaining to the local bulk properties of solar wind, the essential microscopic differences between a wave-particle interacting collisionless gas (see Section III) and a collisional continuum must be noted, particularly in the study of some small-scale phenomena, e.g. the collisionless bow shock transition in front of the magnetosphere.

The essential physics of the hydrodynamic coronal expansion is closely analogous to the flow of a gas through a convergent–divergent nozzle under the influence of pressure drop from the inlet to the exit of the nozzle. The role of constraint influence exerted by the nozzle wall is assumed herein by the solar gravity on the outflowing coronal gas. Following Parker (1958) we idealize the problem by using a steady state, spherically symmetric model in the absence of external magnetic field and by neglecting the viscous effect as well as the thermal radiation transfer of the gas. It is assumed that above a thin coronal base where the solar heat is supplied the only heat transfer is conductive.* The spherically symmetric radial flow is governed by the following system of equations of continuity, momentum, and energy:

$$nur^2 = n_R u_R R^2 \quad \text{(continuity)}, \tag{4.1}$$

$$nmu \frac{du}{dr} = -\frac{dp}{dr} - \frac{GMm}{r^2} \quad \text{(momentum)}, \tag{4.2}$$

$$3nK\mathbf{u} \cdot \nabla T = -2nKT\nabla \cdot \mathbf{u} + \nabla \cdot (k_c \nabla T) \quad \text{(energy)}, \tag{4.3}$$

where n denotes the number density of protons of mass m in a fully ionized coronal (hydrogen) gas, of which the thermal capacity is $3nK$ per unit volume; the pressure is $2nKT$; and the thermal conductivity is k_c. M represents the solar mass, and G the gravitational constant. The subscript R denotes values at the base altitude $r = R$. The above equations are to be supplemented by a thermodynamic polytropic relation between pressure (p) and density (n), namely $p/n^\alpha =$ constant, where the value of the index α is assigned semiempirically ($1 \leq \alpha \leq 5/3$).

The energy equation (4.3), combined with (4.1), becomes

$$\frac{d}{dr}\left(k_c r^2 \frac{dT}{dr}\right) = n_R u_R R^2 \left[3K \frac{dT}{dr} + \frac{2KT}{r^2 u} \frac{d(ur^2)}{dr}\right]. \tag{4.4}$$

The momentum equation (4.2) can be rewritten in the form

$$\frac{2KT}{r^2 u} \frac{d}{dr}(r^2 u) = \frac{d}{dr}\left(\frac{1}{2} mu^2 - \frac{GMm}{r} + 2KT\right). \tag{4.5}$$

Elimination of the quantity $r^2 u$ between (4.4) and (4.5) leads to a resultant equation which is integrable over r. The result is

$$\frac{1}{2} mu^2 - \frac{GMm}{r} + 5KT - \frac{k_c r^2 (dT/dr)}{n_R u_R R^2} = \frac{1}{2} mu_\infty{}^2, \tag{4.6}$$

provided that the temperature falls to zero as r approaches infinity (∞) and $u(\infty) = u_\infty$.

The energy equation (4.6) states that the sum of kinetic energy ($mu^2/2$), gravitational potential energy ($-GMm/r$), thermal energy ($3KT$), work done by pressure (p) through velocity (u) which amounts to $2KT$, and the energy transmitted by thermal conduction is equal to a constant which is the kinetic energy of protons at infinity, namely $mu_\infty{}^2/2$. Hence as far as

* An alternative mechanism of coronal heating, e.g. by wave propagation outward from the sun, will be discussed in Section V.

energy is concerned, the effects of gravitational field and pressure are commutative. The fact that the solar gravitational force acts to constrict the radial expansion of coronal flow in analogy to the cross-sectional area contraction of the converging flow in a Laval nozzle can be further demonstrated. Equation (4.1) can be written as

$$nuA = \text{const.}, \tag{4.7}$$

where the spherical surface area $A = 4\pi r^2$. Differentiation of (4.7) gives

$$\frac{dn}{n} + \frac{du}{u} + \frac{dA}{A} = 0. \tag{4.8}$$

Combining (4.8) and (4.2) to eliminate the variable n, we have

$$\left(1 - \frac{GM}{2a^2 r}\right)\frac{dA}{A} = \left(\frac{u^2}{a^2} - 1\right)\frac{du}{u}, \tag{4.9}$$

where $a^2 = dp/d(mn)$. Note that (4.9) with the gravitational (G) term omitted becomes the classical area–velocity relation of nozzle flow (Liepmann and Roshko, 1957, Chap. 5). It is found that this simple analogy between coronal expansion and Laval nozzle flow is helpful conceptually in studying the solar wind.

To go any further into the study of coronal flow, a supplementary relation between the thermodynamic variables is needed. An outstanding question in contemporary developments of the theory of solar-corona expansion is whether energy is supplied to the quiet corona mainly by thermal conduction outward from a region of active heating at its base, as assumed here, or mainly by wave propagation outward from the sun. In the present approximation the effects of coronal heating and thermal conduction will be represented by a polytropic law relating p and n, i.e. $p/n^{\alpha} = \text{const.}$ or $T = T_0(n/n_0)^{\alpha - 1}$ in which the index α for outward flowing gas is taken to be less than the adiabatic value $\alpha = 5/3$. It has been determined from observed coronal temperature distribution to be $1.1 < \alpha < 1.2$ (Parker, 1963).

Let $U^2 = (1/2)mu^2/2KT_0$, the dimensionless kinetic energy of an atom and $H = GMm/2RKT_0$, the dimensionless gravitational potential of an atom at the reference altitude $r = R$. Substitution of the above polytropic (T, n) relation into the momentum equation (4.2) gives

$$2U - \frac{\alpha U_0^{\alpha-1}}{U^{\alpha}\zeta^{2(\alpha-1)}} = \frac{2\alpha U_0^{\alpha-1}}{U^{\alpha-1}\zeta^{5(\alpha-1)+1}} - \frac{H}{\zeta^2}, \tag{4.10}$$

where $\zeta = r/R$. Integration of (4.10) over ζ gives

$$U^2 + \frac{\alpha}{\alpha - 1}\left(\frac{U_0}{U\zeta^2}\right)^{\alpha - 1} - \frac{H}{\zeta} = U_0^2 - \frac{\alpha}{\alpha - 1} - H = U_1^2. \tag{4.11}$$

The coronal flow velocity $U(\zeta)$, obtained as the solution to (4.11), has two families of curves, one of which is the subsonic branch which has $u < a$ throughout, and the other of which passes through a critical point at $u = a$ and becomes supersonic afterward. The boundary condition at the exit ($r \to \infty$) dictates as to which of the velocity profiles prevails. The above flow description is not unfamiliar to fluid dynamicists who study compressible flows in a convergent–divergent nozzle. It had caused some controversy among astrophysicists (Parker, 1963) as to which of the velocity profiles the solar wind follows.

In order to shed some light on the discussion, let us consider (4.9). If u increases monotonically as A increases, or r increases, the flow is subsonic ($u < a$) when $GM/(a^2 r) > 2$, supersonic ($u > a$) when $GM/(a^2 r) < 2$, and sonic ($u = a$) when $GM/(a^2 r_c) = 2$, where r_c is the critical altitude defined by

$$r_c = GM/2a^2. \tag{4.12}$$

In analogy to the convergent–divergent flow, the coronal expansion has a solar gravitational throat at $r = r_c$. It is therefore the solar gravity that permits the pressure buildup in the lower corona, thus causing the acceleration of corona gas to sonic velocity at $r = r_c$, which is equal to about two or three solar radii* from the center of the sun. Beyond that, the solar corona expands and accelerates to hypersonic velocities. It is of interest to note that had the corona temperature been higher than about 10^6 °K, solar gravity would not be adequate to "throat" the flow to sonic velocity; hence solar wind would be subsonic.

V. A Wave-Pump Problem

A glance at the temperature profile of the chromosphere and corona of the sun (see Fig. 1), which shows that the observed temperature distribution has a steady increase with altitude (measured outward from the solar surface) beyond the photosphere, would bring about a natural question: Considering the fact that the energy source for heating the chromosphere and corona is located beneath the solar surface, how does the solar energy transfer upward against the temperature rise? Thermal conduction is not possible; thermal radiation would not be sufficient. It is noted from the second law of thermodynamics that energy transfer from a cold to a hot body is possible only when work is done, as for example in a heat pump. The energy supplied to the solar corona to maintain its million degree

* Corresponding to the visible solar surface of photosphere.

temperature and supersonic expansion (see Section IV) is believed to come from the dissipation of hydrodynamic and hydromagnetic waves generated by the convective motions in the ionization zone beneath the photosphere, and by photospheric granules, and by the spicules in the chromosphere. The idea is based on the estimations that there is sufficient energy available in the above-mentioned wave-producing regions in the photosphere. The propagation of these waves which carry energy with them to the upper solar atmosphere where the density is low and the wave energy is dissipated could play an important role for the coronal heating. This is designated herein as the wave-pump problem. The prime candidates for the propagating wave modes in question consist of acoustical waves, internal gravity waves, and hydromagnetic waves. It is obvious that none of these three basic modes is expected to exist in pure form. For instance the acoustical waves are likely to be modified by the gravitational field and the magnetic field. A complete analysis must involve plasma instabilities which causes wave generation, propagation, and dispersion, as well as dissipation of waves, including the nonlinear coupling effects, which are essential with waves of finite amplitude. This would be far beyond the scope of the present study. It is intended here to illustrate the essential physics of a wave-pumping action using a simple relevant wave mode as a sample.

The possible importance of internal gravity waves in the role of wave-pumping for energy transport in coronal heating has been discussed in recent works (Whitaker, 1963; Lighthill, 1965). It has been shown that internal gravity waves are present in copious supply and seem to account for much of the heating of the corona. It is postulated that waves will be generated by motion characteristic of photospheric granules, propagate upward into the exponentially decreasing isothermal* atmosphere, and dissipate in the corona. The plausibility for this role of internal gravity waves will be demonstrated by examination of their dispersion characteristics in the solar atmosphere.

Consider a stationary (stable) atmosphere with uniform temperature and composition in which hydrodynamic disturbances of small amplitudes are developed. It is assumed for simplicity that gravity with a constant gravitational acceleration g is the only external force. Let Z be along the upward direction. The equilibrium atmospheric pressure (p_0) and mass density ($\rho_0 = mn_0$) are related by the hydrostatic equation

$$dp_0/dz = -\rho_0 g, \tag{5.1}$$

which, for an isothermal atmosphere, gives

$$p_0(Z)/p_0(0) = \rho_0(Z)/\rho_0(0) = \exp(-mgz/KT). \tag{5.2}$$

* Chosen for simplicity of the analysis.

It is assumed, without loss of generality, that the wave modes lie in the x, y plane. Let the pressure (p), density (ρ), and temperature (T) of the slightly disturbed atmosphere be represented as $p = p_0 + p_1$, $\rho = \rho_0 + \rho_1$, and $T = T_0 + T_1$ respectively, where the subscript 1 denotes small perturbations such that $p_1/p_0 \ll 1$, $\rho_1/\rho_0 \ll 1$, and $T_1/T_0 \ll 1$. The linearized equations of continuity, momentum (x and z components), and energy governing the above perturbations become (Hines, 1960; Yih, 1965, Chap. 2)

$$\frac{\partial \rho_1}{\partial t} + u_z \frac{d\rho_0}{dz} + \rho_0\left(\frac{\partial u_x}{\partial x} + \frac{\partial u_z}{\partial z}\right) = 0, \tag{5.3}$$

$$\rho_0 \frac{\partial u_x}{\partial t} + \frac{\partial p_1}{\partial x} = 0, \tag{5.4}$$

$$\rho_0 \frac{\partial u_z}{\partial t} + \frac{\partial p_1}{\partial z} + \rho_1 g = 0, \tag{5.5}$$

$$\frac{\partial p_1}{\partial t} + u_z \frac{dp_0}{dz} - a_0^2\left(\frac{\partial \rho_1}{\partial t} + u_z \frac{d\rho_0}{dz}\right) = 0, \tag{5.6}$$

where $a_0^2 = \gamma K T_0/m$. Note that (5.6) expresses the adiabatic energy relation in an inviscid atmosphere. Combining (5.2), (5.3), and (5.6) leads to

$$\frac{\partial p_1}{\partial t} - \rho_0 g u_z + a_0^2 \rho_0\left(\frac{\partial u_x}{\partial x} + \frac{\partial u_z}{\partial z}\right) = 0. \tag{5.7}$$

To investigate the allowable wave modes governed by the above equations, let

$$\frac{u_x}{A_1} = \frac{u_z}{A_2} = \frac{\rho_1}{A_3 \rho_0} = \frac{p_1}{A_4 p_0} = \exp i(\omega t + k_x x + k_z z), \tag{5.8}$$

where A_1, A_2, A_3, and A_4 are constants. Substitution of relations (5.8) into (5.3)–(5.6) gives the dispersion relation of the wave modes of interest:

$$\omega^4 - \omega^2[a_0^2(k_x^2 + k_z^2) + i\gamma g k_z] + g^2 k_x(\gamma - 1) = 0. \tag{5.9}$$

Following Hines (1960), let $ik_z = ik_z' + (2H)^{-1}$, where $H = KT/mg$, the atmospheric scale height, and rewrite (5.9) as

$$\omega^4 - \omega^2\left[a_0^2(k_x^2 + k_z'^2) + \frac{\gamma^2 g^2}{4a_0^2}\right] + g^2(\gamma - 1)k_x^2 = 0, \tag{5.10}$$

the roots of which are

$$\omega^2 = \frac{1}{2}\left\{a_0^2(k_x^2 + k_z'^2) + \frac{\gamma^2 g^2}{4a_0^2} \pm \left[\left(a_0^2 k_x^2 + a_0^2 k_z'^2 + \frac{\gamma^2 g^2}{4a_0^2}\right)^2 - 4g^2(\gamma - 1)k_x^2\right]^{1/2}\right\}. \tag{5.11}$$

It is noted that the root with the upper sign (+) corresponds to acoustic waves (ω_A), and the root with the lower sign (—) to internal gravity waves (ω_I).

Consider the acoustic branch ω_A, which has a minimum frequency $\omega_A = \gamma g/2a_0$ when k_x and k_z are both zero (infinite long wavelengths). In the limit of large wave numbers (short wavelengths)

$$\omega_A = a_0 k[1 + O(\omega_A^2/a^2k^2)], \tag{5.12}$$

where $k^2 = k_x^2 + k_z^2$. Note that the ω_A given by (5.12) corresponds to ordinary acoustic waves. Consider now the internal gravity wave branch (ω_I). It reaches a maximum $\omega_b = (\gamma - 1)^{1/2}(g/a_0)k_x/k$ in the limit of large k_x and k_z. At the longer wavelengths and lower frequencies characteristic of the granule motions in the photosphere, $\omega_I \simeq 2(\gamma - 1)^{1/2} a_0 k_x/\gamma$.

From a painstaking study of the comparative frequency characteristics of the acoustic and the internal gravity waves with reference to the observed granule motions in the photosphere, Whitaker (1963) concluded that the observed granules are gravity waves driven by the convective forces below the photosphere, and suggested that it may be gravity waves rather than acoustic waves which get through to heat the solar atmosphere. He further shows that the dissipation of gravity waves is primarily due to thermal conductivity which increases rapidly with rising temperature. The wave energy is dissipated rather quickly when internal gravity waves enter the corona where the temperature rises above 10^6 °K.

VI. Free Expansion Phenomenon

In gas dynamics, free expansion refers to particle streaming from a source into a vacuum with or without the presence of an external force field. The problem of a spherical source of charged particles, which entails collective behavior, and hence quasi-continuum characteristics, has been discussed in Section IV as a topic of large-scale expansion flow of solar corona. In cases when the particles are of neutral species, particularly with tenuous source-particle density such that the size of the effusion orifice is much smaller than the source-particle mean free path, methods of free molecular kinetics are needed for their analysis (Liu and Inger, 1960; Narasimha, 1962). There are numerous astrophysical examples of free expansion phenomena, e.g. the expansion of a particle cloud in the interplanetary (or interstellar) space or the motion of dust particles in a cometary (or meteor) tail (Liu, 1970). The accretion of interstellar particles by a moving planet on account of the planetary gravitational field can be considered as a negative free expansion (or fall-in). It involves the capture of mass by an astronomical body that has significant gravitational field of its own (Spiegel, 1970). The mechanics of accretion is similar to that of free expansion except that the particles move inward.

Consider a balloon whose size is much smaller than the mean free path of its enclosed gas. Collisions with the inner surface of the balloon ensure that the gas has a Maxwellian distribution. When the balloon is burst, the enclosed gas expands freely into a vacuum or a region with nearly zero pressure. The motion of the effusing molecules can be used to simulate that of the freely expanding cloud particles or that of the cometary dust particles emitted by a comet in the interplanetary space. The distribution function $f(\mathbf{x}, \mathbf{v}, t)$ of the freely expanding particles from the erupted balloon source can be determined as the solution of a mathematical initial value problem of the collisionless Boltzmann equation (2.9). This is most conveniently treated by the method of characteristics (see Section II,C).

It is assumed that the initial particle distribution at $t=0$ is given by

$$f(\mathbf{x}, \mathbf{v}, 0) = f_0(\mathbf{x}, \mathbf{v}), \tag{6.1}$$

where f_0 can be an arbitrary function of $\mathbf{x}$ and $\mathbf{v}$. The particle distribution $f(\mathbf{x}, \mathbf{v}, t)$ at time t can be written immediately in terms of its initial distribution f_0, because its variables at time t are all prescribed by the equations of characteristics (2.10) with the external force terms neglected:

$$f(\mathbf{x}, \mathbf{v}, t) = f_0(\mathbf{x} - \mathbf{v}t, \mathbf{v}). \tag{6.2}$$

Particle density $n(\mathbf{x}, t)$ can be determined by integrating $f(\mathbf{x}, \mathbf{v}, t)$ over the velocity space:

$$n(\mathbf{x}, t) = \int f_0(\mathbf{x}, \mathbf{v}t, \mathbf{v})\, d\mathbf{v} = \frac{1}{t^3} \int f_0\left(\mathbf{x}', \frac{\mathbf{x} - \mathbf{x}'}{t}\right) d\mathbf{x}' \tag{6.3}$$

where $\mathbf{x}' = \mathbf{x} - \mathbf{v}t$. Other macroscopic flow quantities of interest which are defined as (velocity) moments of distribution (6.2) can be evaluated accordingly. For example, the flow velocity is

$$\begin{aligned}\mathbf{u}(\mathbf{x}, t) &= \frac{1}{n(\mathbf{x}, t)} \int \mathbf{v} f(\mathbf{x}, \mathbf{v}, t)\, d\mathbf{v} \\ &= \frac{1}{n(\mathbf{x}, t)} \int \left(\frac{\mathbf{x} - \mathbf{x}'}{t}\right) f_0\left(\mathbf{x}', \frac{\mathbf{x} - \mathbf{x}'}{t}\right) d\mathbf{x}' \\ &= \frac{\mathbf{x}}{t} - \frac{1}{nt^4} \int \mathbf{x}' f_0\left(\mathbf{x}', \frac{\mathbf{x} - \mathbf{x}'}{t}\right) d\mathbf{x}'. \end{aligned} \tag{6.4}$$

Any other flow quantity can be similarly calculated as some transformed moment of the initial distribution function.

In a kinetic study of free molecular expansion of a gas cloud of arbitrary configuration, Molmud (1960) draws the formal analogy between the mathematical formulation (6.3) and the solution of the heat equation in terms of Green's function (Carslaw and Jaeger, 1947, Chap. 14). Thus it is possible to use the known solutions of heat equation of various initial geometric configurations (Carslaw and Jaeger, 1947, Chap. 14) to evaluate the free expansion of a cloud of similar initial configuration. It is of interest to examine this formal analogy of Molmud (1960) farther.

Let us differentiate the integral (6.3) with respect to $\mathbf{x}$, obtaining

$$\frac{\partial n(\mathbf{x}, t)}{\partial x} = \frac{1}{t^3} \int \frac{1}{t} \frac{\partial f_0(\mathbf{x}', \xi)}{\partial \xi} \, d\mathbf{x}', \tag{6.5}$$

where $\xi = \xi(\mathbf{x}') = (\mathbf{x} - \mathbf{x}')/t$. Comparison of (6.4) with (6.5) reveals that $\partial u/\partial \mathbf{x}$ is proportional to $n\mathbf{u}$ if and only if

$$\partial f_0(\mathbf{x}', \xi)/\partial \xi = c\xi f_0, \tag{6.6}$$

where C is a constant. It is hence concluded that f_0 must be given as

$$f_0 = g(x) e^{-\beta v^2}, \tag{6.7}$$

which is a Gaussian distribution, to which the Maxwellian molecular distribution belongs. Under the condition (6.7) the particle flux ($n\mathbf{u}$) is proportional to the density gradient ($\partial n/\partial \mathbf{x}$),

$$n\mathbf{u} = -(t/2\beta) \partial n/\partial \mathbf{x}, \tag{6.8}$$

and the flow is irrotational. Substitution of relation (6.8) into the continuity equation

$$\frac{\partial n}{\partial t} + \nabla \cdot (n\mathbf{u}) = 0$$

leads to

$$\frac{\partial n}{\partial t} = \frac{t}{2\beta} \nabla^2 n, \tag{6.9}$$

which is a diffusion equation with a time-dependent diffusion coefficient (Narasimha, 1962). This establishes the mathematical connection between the "free molecular diffusion" represented by (6.9) and the classical heat (ordinary diffusion) equation (Carslaw and Jaeger, 1947, Chap. 1).

It is important to note that the physical mechanism of the ordinary diffusion relies on particle collisions to randomize the particle paths, i.e. to

Gaussianize the particle distribution. On the other hand the "free molecular diffusion" is a collisionless process starting with a Gaussian distribution, with the particles merely going through a random migration.

The treatment of a one-dimensional free expansion in the presence of an external (gravitational) force field does not present much mathematical difficulty and has been given by Keller (1948). With a spherical source in a radial gravitational force field, the expansion with first-order collision effect taken into account becomes the famous problem of planetary atmospheric escape which has not been completely understood. A preliminary kinetic theory of planetary atmospheric escape in the spirit of almost-free molecular flow has been presented by Hays and Liu (1965).

VII. A Generalized Free Expansion Problem

Consider the dust particles which are emitted from a cometary nucleus at speeds much less than the cometary molecular thermal speeds. The expanding gas of the cometary nucleus accelerates the dust particles outward by drag forces. The accelerated dust particles eventually reach velocities of the order of the asymptotic expansion velocity of the gas which is of the order of kilometers per second at radial distances of the order 10^3 km from the nucleus. This is the dust-tail model of Finson and Probstein (1968) who treat the cometary dust motion as a hypersonic collision-free source flow, taking into account solar gravity and radiation pressure acting on the dust particles, thus placing their analysis within the framework of free expansion. A fluid-mechanical study of a cometary (or meteor) tail can be made at various levels of approximation to the effect of dust-gas collisions. Finson and Probstein (1968) assumed the particles to be collisionless, and Liu (1970) assumed them to be collision-dominated. It is found that a general probabilistic method can be given to treat the problem of free expansion of particles which may have velocity-dependent retardation.

The kinetic theory of free expansion (Narasimha, 1962; Keller, 1948) can be viewed alternatively in the light of random function analysis as the determination of the distribution function of new random variables related functionally to the distribution function of the random variables representing the initial conditions (Chandrasekhar, 1943).

The principle of the present stochastic method can be briefly stated. Let $\mathbf{x}(t)$ describe the motion of a cloud particle with t as time. The function $\mathbf{x}(t)$ is a random variable, where random elements are introduced only through initial conditions. Once started, at $t=0$, the random variable $\mathbf{x}(t)$ develops with time according to a deterministic law described by the Newtonian equation of motion of a particle. If the initial conditions $\mathbf{x}(0)$

and $d\mathbf{x}(t)/dt|_{t=0}$ lie initially within a given set S with a specified probability attached to this set, then as time progresses, the values of $\mathbf{x}(t)$ and $d\mathbf{x}(t)/dt$ will lie at each t in the set S_t which has been obtained by transferring all points of S according to the motion. As the motion itself is deterministic, it is evident that the probability of finding $\mathbf{x}(t)$ and $\mathbf{x}(t)/dt$ in S_t should be the same as that of finding $\mathbf{x}(0)$ and $d\mathbf{x}(t)/dt|_{t=0}$ in S. In other words, both sets of S_t and S have the same probability. The above statement on the determination of distribution functions for solution of equations with random initial condition can be generalized to the n-dimensional random vector.

Let $\mathbf{x}=(x_1, \ldots, x_n)$ be an n-dimensional random vector of n random variables x_i $(i=1, \ldots, n)$ defined on some probability space. Let g be a transformation of the random vector $\mathbf{x}$ into a new (also n-dimensional) random vector $\mathbf{y}=(y_1, \ldots, y_n)$ defined on the same probability space. The mapping of g is deterministic. The problem is to determine the distribution of the transformed vector $\mathbf{y}=g(\mathbf{x})$ in terms of the distribution of the original vector $\mathbf{x}$. The basic principle of interest here is that an event in probability space should have the same probability irrespective of whether it is described in terms of the random vector $\mathbf{x}$ or the random vector $\mathbf{y}$, both vectors $\mathbf{x}$ and $\mathbf{y}$ being related by a deterministic transformation g. It is assumed that g be defined by a set of n relations

$$y_i=g_i(x_1, \ldots, x_n), \qquad i=1, \ldots, n$$

where the functions g_i are continuous, have continuous partial derivatives, and provide a one-to-one correspondence between $\mathbf{x}$ and $\mathbf{y}$. Hence the inverse functions

$$x_i=h_i(y_1, \ldots, y_n), \qquad i=1, \ldots, n$$

characterizing the inverse transformation $g^{-1}=h$ exist, are also one-to-one, and have continuous partial derivatives. Under the conditions stated above, the joint density of a transformed random vector $(y_1, \ldots, y_n)$ is

$$f_y(y_1, y_2, \ldots, y_n)=f_x(x_1, x_2, \ldots, x_n)/|J(x_1, x_2, \ldots, x_n)|, \tag{7.1}$$

where

$$J(x_1, x_2, \ldots, x_n)=\begin{vmatrix} \dfrac{\partial g_1}{\partial x_1} & \cdots & \dfrac{\partial g_1}{\partial x_n} \\ \vdots & & \vdots \\ \dfrac{\partial g_n}{\partial x_1} & \cdots & \dfrac{\partial g_n}{\partial x_n} \end{vmatrix} \equiv \frac{\partial(g_1, \ldots, g_n)}{\partial(x_1, \ldots, x_n)}$$

is the Jacobian of transformation.

The above principle will be applied to study the expansion of a particle cloud in a resisting medium. It is assumed that the initial distribution of the cloud particles is Maxwellian, although this is not necessary so far as the method is concerned. The equations of motion of a cloud particle of mass m can be written as

$$\partial^2\mathbf{x}/\partial t^2 = -\nu\,\partial\mathbf{x}/\partial t, \qquad \partial\mathbf{x}/\partial t = \mathbf{v}, \tag{7.2}$$

which state that resistance to the particle motion follows Stoke's law with $\nu = 6\pi\,\mu d/m$, μ being the viscosity and d the diameter of the particle. The solution of (7.2) is

$$x - x_0 = (\mathbf{v}_0/\nu)(1 - e^{-\nu t}), \qquad \mathbf{v}/\mathbf{v}_0 = e^{-\nu t}. \tag{7.3}$$

The application of relation (7.1) to the present problem gives

$$n\,\frac{\partial(x_1, x_2, x_3)}{\partial(v_{01}, v_{02}, v_{03})} = n_0(\beta/\pi)^{3/2}\exp[-\beta(v_{01}^2 + v_{02}^2 + v_{03}^2)] \tag{7.4}$$

which, together with solutions (7.3) to eliminate $\mathbf{v}_0$, leads to the density distribution

$$n(\mathbf{x}, t) = n_0\nu^3\left(\frac{\beta}{\pi}\right)^{3/2}(1 - e^{-\nu t})^{-3}\exp\left[-\frac{\beta\nu^2}{(1 - e^{-\nu t})^2}\sum_{i=1}^{3}(x_i - x_{0i})^2\right]. \tag{7.5}$$

This is reducible to a special solution of free expansion (Narasimha, 1962) by letting $\nu = 0$.

When the initial particle cloud occupies a domain V prior to the start of expansion, the corresponding density and mean particle velocity can be evaluated as the moments of $n(\mathbf{x}, t)$ given by (7.5):

$$\begin{aligned} n'(\mathbf{x}, t) &= \int_V n(\mathbf{x}, \mathbf{x}_0, t)\,d\mathbf{x}_0 \qquad (d\mathbf{x}_0 = dx_1, dx_2, dx_3) \\ &= n_0\nu^3\left(\frac{\beta}{\pi}\right)^{3/2}\int(1 - e^{-\nu t})^{-3}\exp\left[-\frac{\beta\nu^2}{(1 - e^{-\nu t})^2}\sum_{i=1}^{3}(x_i - x_{0i})^2\right]d\mathbf{x}_0, \end{aligned} \tag{7.6}$$

$$\begin{aligned} u &= \frac{1}{n'}\int_V n\mathbf{v}\,d\mathbf{x}_0 \\ &= n_0\nu^4\left(\frac{\beta}{\pi}\right)^{3/2}\frac{e^{-\nu t}}{n'}\int_V(\mathbf{x} - \mathbf{x}_0)(1 - e^{-\nu t})^{-4} \\ &\qquad \exp\left[-\frac{\beta\nu^2}{(1 - e^{-\nu t})^2}\sum_{i=1}^{3}(x_i - x_{0i})^2\right]d\mathbf{x}_0. \end{aligned} \tag{7.7}$$

These results agree with those of Levin (1967), who uses the Lagrangian method of hydrodynamics (Kochin *et al.*, 1949, p. 15).

The above results can be used with advantage in the study of the expansion of a dust cloud emitted from a cometary nucleus when the resistance to the motion of a dust particle is not negligible.

The above method is found possible even when the resistance is proportional to the square of the particle velocity, at least in the one-dimensional case. Consider particles falling under gravitational acceleration (g) along the x axis. For this motion $v(t)=dx/dt$, $v(0)=v_0$, and

$$\frac{dv(t)}{dt}+\frac{D}{m}v^2(t)=g, \tag{7.8}$$

which is a special case of the Riccati equation, and has the solution

$$v(t)=\left(\frac{gm}{D}\right)^{1/2}\frac{(gm/D)^{1/2}\tanh\alpha t+v_0}{(gm/D)^{1/2}+v_0\tanh\alpha t}, \qquad (0\le t\le\infty) \tag{7.9}$$

where $\alpha=(gD/m)^{1/2}\equiv v(\infty)$. The inverse relation is

$$v_0=\left(\frac{gD}{m}\right)^{1/2}\frac{(gD/m)^{1/2}\tanh\alpha t-v}{v\tanh\alpha t-(gD/m)^{1/2}}, \qquad (0\le t<\infty). \tag{7.10}$$

Hence

$$\frac{dv_0}{dv}=\left(\frac{gD}{m}\right)^{1/2}\frac{1-\tanh^2\alpha t}{[(gD/m)^{1/2}-v\tanh\alpha t]^2}\ge 0.$$

Let the random variable v_0 have initial distribution $f_0(v_0)$ for $v_0>0$ and $f_0(v_0)=0$ for $v_0\le 0$. By relation (7.1), the distribution of particles at t after the expansion is

$$f(v,t)=f_0(v_0)\,dv_0/dv. \tag{7.11}$$

Consider a simple case of uniform initial particle distribution

$$f_0(v_0)=\begin{cases}k_1 & \text{if} \quad 0<v_0\le k_1^{-1},\\ 0 & \text{otherwise.}\end{cases}$$

The mean particle velocity at t is

$$\begin{aligned}v(t)&=k_1\left(\frac{gm}{D}\right)^{1/2}\int_0^{k_1^{-1}}\frac{(gm/D)^{1/2}\tanh\alpha t+v_0}{(gm/D)^{1/2}+v_0\tanh\alpha t}\,dv_0\\ &=\frac{gm/D}{\sinh^2\alpha t}\left[\frac{\sinh 2\alpha t}{2(gm/D)^{1/2}}-\ln\frac{(gm/D)^{1/2}+k_1^{-1}\tanh\alpha t}{(gm/D)^{1/2}}\right],\end{aligned} \tag{7.12}$$

with $v(0)=k_1^{-1}/2$ and $v(\infty)=(gm/D)^{1/2}$.

In many free expansions of astrophysical interest, the statistical data on the particle motions are usually not so complete such that a distribution function can be determined or even inferred. Instead, sole limited information on the fluctuation characteristics of the particles may be available and it is often of interest to relate the statistical behavior of particles to that prior to the start of cloud expansion. Let the initial particle motions be denoted by a random process $\mathbf{x}(t)$ and the expanded particle motions $\mathbf{y}(t)$ such that

$$\mathbf{y}(t) = T\mathbf{x}(t), \tag{7.13}$$

where the operator T designates the rule of correspondence between the random functions $\mathbf{x}(t)$ and $\mathbf{y}(t)$. The process $\mathbf{y}(t)$ can be considered as the output of a system with input $\mathbf{x}(t)$. (It is noted that the response of systems to stochastic inputs is the same as the response to ordinary functions. The randomness manifests itself on the selection of a particular input, and not on the nature of this input or the way the system responds to it.) In the limited cases of memoryless time-invariant and linear systems, a considerable volume of results on linear transformation of random processes is available, e.g. the relations pertaining to correlations and frequency spectra of inputs and outputs (Liepmann, 1952).

VIII. Concluding Remarks

In spite of the important role being played by the dynamics of gases under cosmic condition in space research, there is apparently not an introduction on the subject suitable for fluid dynamicists who are interested in the adjoining field of astrophysical research. To achieve this, complex astrophysical phenomena have been avoided in the presentation as far as possible; emphasis has been placed on the features unique to a cosmic plasma which might be unfamiliar to fluid dynamicists. These include the fluidlike properties of an interplanetary plasma medium and the effective wave transport of mechanical energy under interplanetary conditions. The close fluid-dynamical analogy between coronal expansion and Laval nozzle flow (see Section IV) cannot fail to present to fluid dynamicists a broader outlook of their discipline. On this particular point, it is of interest to note the similarity in the mathematical treatment of a disturbing gravitating planet in the interplanetary space (Spiegel, 1970) and a diagnostic probe in a collisionless plasma (Liu, 1969). Another case in point is the similar nature between the collective behavior of stars in the presence of a gravitational field (Lynden–Bell, 1967) and that of plasma particles in a Coulomb field (see Section III).

The individual problems presented herein are idealized sufficiently to make them tractable with the methods of formal analysis, yet with the essential physics of interest retained in this idealization. Nonetheless the results of these severely idealized models serve only to illustrate the physical phenomena and are not to be regarded as definite and quantitative. The goal of the present discussion of selected topics is to put together some representative interplanetary problems which are examples of either continuum or free particle flows.

Some of the unsolved problems related to present topics are now briefly discussed. Regarding the theory of large-scale expansion of the solar corona, the outstanding question is whether the energy supply to the quiet corona is primarily a thermal conduction effect, as postulated in the formulation of Section IV, or a wave propagation effect which is discussed in Section V. In the latter, the specific wave mode of primary importance for such energy transfer in the corona has not been clearly identified with any of the likely candidates of acoustic, internal gravity, and hydromagnetic waves, among others. An elimination process based on the wave generation, dispersion, and propagation characteristics relative to the observed corona would be helpful in this regard.

A very challenging problem pertaining to the microscopic interaction between solar wind and cometary atmosphere is worthy of notice. It has been shown from observations on comets that the copious ionization in the cometary atmosphere, apparently caused by its interaction with the solar wind, cannot be satisfactorily explained by any and all of the contemporarily suggested ionization processes, e.g. the charge-transfer model (Biermann, 1951), the magnetized-plasma-interaction model (Alfven, 1960), and the shock-energized-electron model (Axford, 1964), although the modulation of solar wind by the bow shock in front of a comet appears to have some bearing on the cometary ionization in question. The physics of interaction between a cometary atmosphere and the solar wind is also involved when a nonmagnetic planet, such as Mars and Venus, meets the solar wind. The solution of the cometary problem would have an important bearing on the understanding of planetary ionospheric physics.

ACKNOWLEDGMENT

The present work has been supported by a NASA research Grant NGR 23-005-094.

REFERENCES

Alfven, H. (1960). *Rev. Mod. Phys.* **32**, 710.

Allen, C. W. (1963). "Astrophysical Quantities," 2nd ed. Oxford Univ. Press (Athlone), London and New York.

AXFORD, W. I. (1962). *J. Geophys. Res.* **67**, 3791.

AXFORD, W. I. (1964). *Planet. Space Sci.* **12**, 719.

BATCHELOR, G. K. (1953). "Homogeneous Turbulence." Cambridge Univ. Press, London and New York.

BEHR, A., and SIEDENTOPF, H. (1953). *Z. Astrophys.* **32**, 19.

BETHE, H. A. (1939). *Phys. Rev.* **55**, 434.

BIERMANN, L. (1951). *Z. Astrophys.* **29**, 274.

BIERMANN, L., BROSOWSKI, B., and SCHMIDT, H. U. (1957). *Sol. Phys.* **1**, 254.

BLACKWELL, D. E., and INGHAM, M. F. (1961). *Mon. Notic. Roy. Astron. Soc.* **122**, 129.

BURGERS, J. M. (1969). "Flow Equations for Composite Gases." Academic Press, New York.

CARSLAW, H. S., and JAEGER, J. C. (1947). "Conduction of Heat in Solids." Oxford Univ. Press, London and New York.

CHANDRASEKHAR, S. (1942). "Principles of Stellar Dynamics." Univ. of Chicago Press, Chicago, Illinois.

CHANDRASEKHAR, S. (1943). *Rev. Mod. Phys.* **15**, 1.

CHAPMAN, S., and COWLING, T. G. (1970). "Mathematical Theory of Nonuniform Gases," 3rd ed. Cambridge Univ. Press, London and New York.

CHAPMAN, S., and FERRARO, V. C. A. (1931). *Terr. Magn. Atmos. Elec.* **36**, 77 and 171.

DRUMMOND, W. E., and PINES, D. (1962). *Nucl. Fusion, Suppl.* Part 3, p. 1049.

FINSON, M. L., and PROBSTEIN, R. F. (1968). *Astrophys. J.* **154**, 327.

FRIED, B. D., and GOULD, R. W. (1961). *Phys. Fluids* **4**, 139.

GINZBURG, V. L. (1964). "Propagation of Electromagnetic Waves in Plasmas." Pergamon, Oxford.

GOLD, T. (1955). *In* "Gas Dynamics of Cosmic Clouds" (H. C. Van de Hulst and J. M. Burgers, eds.). North-Holland Publ., Amsterdam.

HARRIS, E. G. (1969). *In* "Advance in Plasma Physics" (A. Simon and W. B. Thompson, eds.), Vol. 3, p. 175. Wiley (Interscience), New York.

HAYS, P. B., and LIU, V. C. (1965). *Planet. Space Sci.* **13**, 1185.

HINES, C. O. (1960). *Can. J. Phys.* **38**, 1441.

HUNDHAUSEN, A. J. (1968). *Space Sci. Rev.* **8**, 690.

HUNG, R. J., and LIU, V. C. (1971). *Phys. Fluids* **14**, 2061.

KADOMTSEV, B. B. (1965). "Plasma Turbulence." Academic Press, New York.

KELLER, J. B. (1948). *Commun. Pure Appl. Math.* **1**, 275.

KELLOGG, P. (1962). *J. Geophys. Res.* **67**, 3805.

KOCHIN, N. E., KIBEL, I. A., and ROZE, N. V. (1949). "Theoretical Hydromechanics," Vol. I. Wiley (Interscience), New York.

KRALL, N. A. (1968). *In* "Advance in Plasma Physics" (A. Simon and W. B. Thompson, eds.), Vol. 2, p. 153. Wiley (Interscience), New York.

LANDAU, L. D. (1946). *J. Phys. (Moscow)* **10**, 25.

LEVIN, V. A. (1967). *J. Appl. Math. Mech. (USSR)* **30**, 736.

LIEPMANN, H. W. (1952). *J. Appl. Math. Phys.* **3**, 321.

LIEPMANN, H. W., and ROSHKO, A. (1957). "Elements of Gasdynamics." Wiley, New York.

LIGHTHILL, M. J. (1965). *In* "5th Symposium on Cosmic Gas Dynamics" (R. N. Thomas, ed.), p. 429. Academic Press, New York.

LIU, V. C. (1965). *Nature (London)* **208**, 883.

LIU, V. C. (1969). *Space Sci. Rev.* **9**, 423.

LIU, V. C. (1970). *Phys. Fluids* **13**, 62.

LIU, V. C. and INGER, G. R. (1960). *J. Aerosp. Sci.* **27**, 875.

Lynden-Bell, D. (1967). *In* "Relativity and Astrophysics" (J. Elhers, ed.), Lect. Appl. Math., p. 131, Amer. Math. Soc., Providence, Rhode Island.
Malmberg, J. H., and Wharton, C. B. (1964). *Phys. Rev. Lett.* **13**, 184.
Molmud, P. (1960). *Phys. Fluids* **3**, 362.
Montgomery, D. C., and Tidman, D. A. (1964). "Plasma Kinetic Theory." McGraw-Hill, New York.
Narasimha, R. (1962). *J. Fluid Mech.* **12**, 294.
Parker, E. N. (1958). *Astrophys. J.* **128**, 664.
Parker, E. N. (1962). *Space Sci. Rev.* **1**, 69.
Parker, E. N. (1963). "Interplanetary Dynamic Processes." Wiley (Interscience), New York.
Rossi, B., and Olbert, S. (1970). "Physics of Space." McGraw-Hill, New York.
Shabansky, V. P. (1968). *Space Sci. Rev.* **8**, 688.
Spiegel, E. A. (1970). *In* "Interstellar Gas Dynamics" (H. Habing, ed.). Reidel, Dordrecht.
Stix, T. H. (1962). "Theory of Plasma Waves." McGraw-Hill, New York.
Sturrock, P. A., and Hartle, R. E. (1966). *Phys. Rev. Lett.* **16**, 628.
Tidman, D. A., and Krall, N. A. (1971). "Shock Waves in Collisionless Plasma." Wiley, New York.
Vedenov, A. A., Velikhov, E. P., and Sagdeev, R. Z. (1962). *Nucl. Fusion, Suppl.* Part 2, p. 465.
Whitaker, W. (1963). *Astrophys. J.* **137**, 914.
Yih, C. S. (1965). "Dynamics of Nonhomogeneous Fluids." Macmillan, New York.

Author Index

Numbers in italics refer to the pages on which the complete references are listed.

M

N

O

P

R

S

Y

Z

Subject Index

T

V

W

Y

Z